HANDBOOK OF
DEEP LEARNING IN
BIOMEDICAL ENGINEERING
AND HEALTH INFORMATICS

Biomedical Engineering: Techniques and Applications

HANDBOOK OF DEEP LEARNING IN BIOMEDICAL ENGINEERING AND HEALTH INFORMATICS

Edited by
E. Golden Julie, PhD
Y. Harold Robinson, PhD
S. M. Jaisakthi, PhD

APPLE ACADEMIC PRESS

First edition published 2022

Apple Academic Press Inc.
1265 Goldenrod Circle, NE,
Palm Bay, FL 32905 USA

4164 Lakeshore Road, Burlington,
ON, L7L 1A4 Canada

CRC Press
6000 Broken Sound Parkway NW,
Suite 300, Boca Raton, FL 33487-2742 USA

2 Park Square, Milton Park,
Abingdon, Oxon, OX14 4RN UK

© 2022 Apple Academic Press, Inc.

Apple Academic Press exclusively co-publishes with CRC Press, an imprint of Taylor & Francis Group, LLC

Library and Archives Canada Cataloguing in Publication

Title: Handbook of deep learning in biomedical engineering and health informatics / edited by E. Golden Julie, PhD, Y. Harold Robinson, PhD, S.M. Jaisakthi, PhD.

Names: Julie, E. Golden, 1984- editor. | Robinson, Y. Harold, editor | Jaisakthi, S. M., editor.

Series: Biomedical engineering (Apple Academic Press)

Description: First edition. | Series statement: Biomedical engineering: techniques and applications | Includes bibliographical references and index.

Identifiers: Canadiana (print) 20210143177 | Canadiana (ebook) 20210143479 | ISBN 9781771889988 (hardcover) | ISBN 9781774638170 (softcover) | ISBN 9781003144694 (ebook)

Subjects: LCSH: Diagnostic imaging. | LCSH: Artificial intelligence. | LCSH: Medical informatics. | LCSH: Biomedical engineering.

Classification: LCC RC78.7.D53 H36 2021 | DDC 616.07/54—dc23

Library of Congress Cataloging-in-Publication Data

Names: Julie, E. Golden, 1984- editor. | Robinson, Y. Harold, editor | Jaisakthi, S. M., editor.

Title: Handbook of deep learning in biomedical engineering and health informatics / edited by E. Golden Julie, Y Harold Robinson, S. M. Jaisakthi

Other titles: Biomedical engineering (Apple Academic Press)

Description: First edition. | Palm Bay, FL : Apple Academic Press, 2021. | Series: Biomedical engineering: techniques and applications | Includes bibliographical references and index. | Summary: "This new volume discusses state-of-the-art deep learning techniques and approaches that can be applied in biomedical systems and health informatics. Deep learning in the biomedical field is an effective method of collecting and analyzing data that can be used for the accurate diagnosis of disease. This volume delves into a variety of applications, techniques, algorithms, platforms, and tools used in this area, such as image segmentation, classification, registration, and computer-aided analysis. The editors proceed on the principle that accurate diagnosis of disease depends on image acquisition and interpretation. There are many methods to get high resolution radiological images, but we are still lacking in automated image interpretation. Currently deep learning techniques are providing a feasible solution for automatic diagnosis of disease with good accuracy. Analyzing clinical data using deep learning techniques enables clinicians to diagnose diseases at an early stage and treat the patients more effectively. Chapters explore such approaches as deep learning algorithms, convolutional neural networks and recurrent neural network architecture, image stitching techniques, deep RNN architectures, and more. The volume also depicts how deep learning techniques can be applied for medical diagnostics of several specific health scenarios, such as cancer, COVID-19, acute neurocutaneous syndrome, cardiovascular and neuro diseases, skin lesions and skin cancer, etc. Key features: Introduces important recent technological advancements in the field Describes the various techniques, platforms, and tools used in biomedical deep learning systems Includes informative case studies that help to explain the new technologies Handbook of Deep Learning in Biomedical Engineering and Health Informatics provides a thorough exploration of biomedical systems applied with deep learning techniques and will provide valuable information for researchers, medical and industry practitioners, academicians, and students"-- Provided by publisher.

Identifiers: LCCN 2021008116 (print) | LCCN 2021008117 (ebook) | ISBN 9781771889988 (hbk) | ISBN 9781774638170 (pbk) | ISBN 9781003144694 (ebk)

Subjects: MESH: Medical Informatics | Deep Learning | Biomedical Engineering

Classification: LCC R855.3 (print) | LCC R855.3 (ebook) | NLM W 26.5 | DDC 610.285--dc23

LC record available at https://lccn.loc.gov/2021008116

LC ebook record available at https://lccn.loc.gov/2021008117

ISBN: 978-1-77188-998-8 (hbk)
ISBN: 978-1-77463-817-0 (pbk)
ISBN: 978-1-00314-469-4 (ebk)

ABOUT THE BOOK SERIES: BIOMEDICAL ENGINEERING: TECHNIQUES AND APPLICATIONS

This new book series aims to cover important research issues and concepts of the biomedical engineering progress in alignment with the latest technologies and applications. The books in the series include chapters on the recent research developments in the field of biomedical engineering. The series explores various real-time/offline medical applications that directly or indirectly rely on medical and information technology. Books in the series include case studies in the fields of medical science, i.e., biomedical engineering, medical information security, interdisciplinary tools along with modern tools, and technologies used.

Coverage & Approach
- In-depth information about biomedical engineering along with applications.
- Technical approaches in solving real-time health problems
- Practical solutions through case studies in biomedical data
- Health and medical data collection, monitoring, and security

The editors welcome book chapters and book proposals on all topics in the biomedical engineering and associated domains, including Big Data, IoT, ML, and emerging trends and research opportunities.

Book Series Editors:
Raghvendra Kumar, PhD
Associate Professor, Computer Science & Engineering Department,
GIET University, India
Email: raghvendraagrawal7@gmail.com

Vijender Kumar Solanki, PhD
Associate Professor, Department of CSE, CMR Institute of Technology
(Autonomous), Hyderabad, India
Email: spesinfo@yahoo.com
Noor Zaman, PhD

School of Computing and Information Technology, Taylor's University, Selangor, Malaysia
Email: noorzaman650@hotmail.com

Brojo Kishore Mishra, PhD
Professor, Department of CSE, School of Engineering, GIET University, Gunupur, Osidha, India
Email: bkmishra@giet.edu

FORTHCOMING BOOKS IN THE SERIES

The Congruence of IoT in Biomedical Engineering: An Emerging Field of Research in the Arena of Modern Technology
Editors: Sushree Bibhuprada B. Priyadarshini, Rohit Sharma, Devendra Kumar Sharma, and Korhan Cengiz

Handbook of Artificial Intelligence in Biomedical Engineering
Editors: Saravanan Krishnan, Ramesh Kesavan, and B. Surendiran

Handbook of Deep Learning in Biomedical Engineering and Health Informatics
Editors: E. Golden Julie, Y. Harold Robinson, and S. M. Jaisakthi

Biomedical Devices for Different Health Applications
Editors: Garima Srivastava and Manju Khari

Handbook of Research on Emerging Paradigms for Biomedical and Rehabilitation Engineering
Editors: Manuel Cardona and Cecilia García Cena

High-Performance Medical Image Processing
Editors: Sanjay Saxena and Sudip Paul

The Role of Internet of Things (IoT) in Biomedical Engineering: Present Scenario and Challenges
Editors: Sushree Bibhuprada B. Priyadarshini, Devendra Kumar Sharma, Rohit Sharma, and Korhan Cengiz

ABOUT THE EDITORS

E. Golden Julie, PhD
Senior Assistant Professor, Department of Computer Science and Engineering, Anna University, Regional Campus, Tirunelveli, India

E. Golden Julie, PhD, is currently working as a Senior Assistant Professor in the Department of Computer Science and Engineering, Anna University, Regional Campus, Tirunelveli. She has more than 12 years of experience in teaching. She has published more than 34 papers in various international journals and has presented more than 20 papers at both national and international conferences. She has written 10 book chapters published by Springer and IGI Global. Dr. Julie is an editor for the book, *Successful Implementation and Deployment of IoT Projects in Smart Cities*, published by IGI Global in the Advances in Environmental Engineering and Green Technologies (AEEGT) book series. She is one of the editors for the book, *Handbook of Research on Blockchain Technology: Trend and Technologies*, published by Elsevier. She is acting as a reviewer of many journals, including *Computers and Electrical Engineering* (Elsevier) and *Wireless Personal Communication* (Springer). She has given many guest lectures in various subjects, such as multicore architecture, operating systems, at compiler design in various institutions.

Dr. Julie is a recognized reviewer and translator for National Programme on Technology Enhanced Learning (NPTEL) online (MOOC) courses. She has acted as a jury member at the national level and the international level for IEEE conferences, project fairs, and symposia. Dr. Julie has completed nine NPTEL courses and earned star certificates, topper designations, and sliver certificates. She attended faculty development programs to enhance the knowledge of the student community. Her research area includes wireless sensor ad-hoc networks, soft computing, blockchain, fuzzy logic, neural network, soft computing techniques, clustering, optimized techniques, IoT, and image processing. She is also an active lifetime member of the Indian Society of Technical Education.

Y. Harold Robinson, PhD

School of Information Technology and Engineering,
Vellore Institute of Technology, Vellore, India

Y. Harold Robinson, PhD, is currently working in the School of Information Technology and Engineering at Vellore Institute of Technology, Vellore, India. He has more than 15 years of experience in teaching. He has published more than 50 papers in various international journals and has presented more than 45 papers in both national and international conferences. He has written four book chapters published by Springer and IGI Global. He is an editor for the book, *Successful Implementation and Deployment of IoT Projects in Smart Cities,* published by IGI Global in the Advances in Environmental Engineering and Green Technologies (AEEGT) book series. He is one of the editors for the book, *Handbook of Research on Blockchain Technology: Trend and Technologies,* published by Elsevier. He is acting as a reviewer of many journals, including *Multimedia Tools and Applications and Wireless Personal Communication* (Springer). He has given many guest lectures in various subjects, such as a pointer, operating system, compiler design at various institutions. He has also provided an invited talk at a technical symposium. He has acted as a convener, coordinator, and jury member at national and international IEEE conferences, project fairs, and symposia. He has attended various seminars, workshops, and faculty development programs to enhance knowledge of the student community. His research area includes wireless sensor networks, ad-hoc networks, soft computing, blockchain, IoT, and image processing. He has published research papers in various SCIE journals, including like *Wireless Personal Communication, IEEE Access, Ad Hoc Networks, Mobile Networks and Applications, Journal of Ambient Intelligence and Humanized Computing, Peer-to-Peer Networking and Applications, Journal of Intelligent and Fuzzy Systems, Computer Standards and Interfaces,* and *Earth Science Informatics.* He is also an active lifetime member of the Indian Society of Technical Education.

S. M. Jaisakthi, PhD
Associate Professor, School of Computer Science and Engineering,
Vellore Institute of Technology, India

S. M. Jaisakthi, PhD, is an Associate Professor in the School of Computer Science and Engineering at Vellore Institute of Technology. She received her undergraduate and a doctoral degrees from Anna University, Chennai, India. Dr. Jaisakthi has extensive research experience in machine learning in the area of image processing and medical image analysis. She also has good experience in building deep learning models, including convolutional (CNN) and recurrent neural networks (RNN). She has published many research publications in refereed international journals and in proceedings of international conferences. Currently, she is investigating a project funded by Science and Engineering Research Board worth Rs. 29.75 lakhs.

CONTENTS

CONTRIBUTORS

M. Anly Antony
Assistant Professor in CSE, Sahrdaya College of Engineering and Technology, Kodakara, Thrissur, Kerala, India, E-mail: anlyantony@sahrdaya.ac.in

K. Anusree
Assistant Professor in CSE, Sahrdaya College of Engineering and Technology, Kerala, India, E-mail: anusreek@sahrdaya.ac.in

Chidambaranathan
Associate Professor in MCA, St. Xavier's College, Tirunelveli, Tamil Nadu, India

A. Devi
Assistant Professor, Department of ECE, IFET College of Engineering, Villupuram, Tamil Nadu, India, E-mail: deviarumugam02@gmail.com

B. Devikirubha
School of Computer Science and Engineering, Vellore Institute of Technology, Tamil Nadu, India, E-mail: devikiruba.b2018@vitstudent.ac.in

R. Divya
Assistant Professor in CSE, Sahrdaya College of Engineering and Technology, Kerala, India, E-mail: divyar@sahrdaya.ac.in

S. M. Jaisakthi
Associate Professor, School of Computer Science and Engineering, Vellore Institute of Technology, Tamil Nadu, India, E-mail: jaisakthi.murugaiyan@vit.ac.in

T. Jarin
Department of Electrical and Electronics Engineering, Jyothi Engineering College, Thrissur, Kerala, India, E-mail: jeroever2000@gmail.com

G. Kavya
Professor, Department of ECE, S. A. Engineering College, Chennai, Tamil Nadu, India, E-mail: gkavya2013@gmail.com

G. R. Gnana King
Associate Professor in ECE, Sahrdaya College of Engineering and Technology, Kodakara, Thrissur, Kerala, India, E-mail: gnanaking@sahrdaya.ac.in

Shyam Krishna
Assistant Professor in CSE, Sahrdaya College of Engineering and Technology, Kerala, India, E-mail: shyamkrishna@sahrdaya.ac.in

A. Jayakumar
Associate Professor, Department of Electronics and Communication Engineering, IFET College of Engineering, Villupuram, Tamil Nadu, India, E-mail: jayangce@gmail.com

K. Suresh Kumar
Associate Professor, Department of Electronics and Communication Engineering, IFET College of Engineering, Villupuram, Tamil Nadu, India, E-mail: sureshkumarkskphd@gmail.com

R. Satheesh Kumar
Associate Professor in CSE, Sahrdaya College of Engineering and Technology, Kodakara, Thrissur, Kerala, India, E-mail: satheeshkumar@sahrdaya.ac.in

T. Ananth Kumar
Assistant Professor, Department of Computer Science and Engineering, IFET College of Engineering, Gangarampalaiyam, Tamil Nadu, India, E-mail: ananth.eec@gmail.com

V. Kishore Kumar
Associate Professor, Department of Electronics and Communication Engineering, IFET College of Engineering, Villupuram, Tamil Nadu, India, E-mail: v.kishore882@gmail.com

P. Manjubala
Associate Professor, Department of CSE, IFET College of Engineering, Gengarampalayam, Tamil Nadu, India, E-mail: pkmanju26@gmail.com

V. Milner Paul
Research Scholar, Department of Electrical Engineering, National Institute of Technology, Manipur, India, E-mail: vithayathilmilner@nitmanipur.ac.in

M. Pavithra
Assistant Professor, Department of CSE, IFET College of Engineering, Gangarampalaiyam, Tamil Nadu, India, E-mail: pavimuthu27@gmail.com

S. R. Boselin Prabhu
Department of Electronics and Communication Engineering, Surya Engineering College, Mettukadai, Tamil Nadu, India, E-mail: eben4uever@gmail.com

A. S. Radhamani
Professor, Department of Electronics and Communication Engineering, V. V. College of Engineering, Tamil Nadu, India, E-mail: asradhamani@gmail.com

R. S. Rajesh
Department of Computer Science and Engineering, Manonmaniam Sundaranar University, Tirunelveli, Tamil Nadu, India, E-mail: rsrajesh@msuniv.ac.in

M. Rajeswari
Professor in CSE, Sahrdaya College of Engineering and Technology, Kerala, India, E-mail: rajeswarim@sahrdaya.ac.in

R. Rajmohan
Associate Professor, Department of CSE, IFET College of Engineering, Gangarampalaiyam, Tamil Nadu, India, E-mail: rjmohan89@gmail.com

S. G. Sandhya
Associate Professor, Department of CSE, IFET College of Engineering, Gangarampalaiyam, Tamil Nadu, India, E-mail: sgsandhyadhas@gmail.com

S. Arunmozhi Selvi
Department of Computer Science and Engineering, Manonmaniam Sundaranar University, Tirunelveli, Tamil Nadu, India, E-mail: heyaruna@gmail.com

M. Sowmiya
Assistant Professor of IT Department, M. Kumarasamy College of Engineering, Karur, Tamil Nadu, India, E-mail: sowmiyam.it@mkce.ac.in

S. Sundaresan
Assistant Professor, Department of Electronics and Communication Engineering, SRM TRP Engineering College, Trichy, Tamil Nadu, India, E-mail: sundaresanece91@gmail.com

V. V. Satyanarayana Tallapragada
Associate Professor, Department of ECE, Sree Vidyanikethan Engineering College, Tirupati, Andhra Pradesh, India, E-mail: satya.tvv@gmail.com

M. Julie Therese
Assistant Professor, Department of ECE, Sri Manakula Vinayagar Engineering College, Pondicherry University, Puducherry, Tamil Nadu, India, E-mail: julietherese@smvec.ac.in

C. Thilagavathi
Assistant Professor of IT Department, M. Kumarasamy College of Engineering, Karur, Tamil Nadu, India, E-mail: thilagavathic.it@mkce.ac.in

V. Yuvaraj
Associate Professor in BM, Sahrdaya College of Engineering and Technology, Kodakara, Thrissur, Kerala, India, E-mail: yuvarajv@sahrdaya.ac.in

ABBREVIATIONS

AAM	active accurate model
ABCD	asymmetric, border structure, color difference, and diameter
AD	Alzheimer's disease
ADHD	attention deficit hyperactivity disorder
ADNI	Alzheimer's disease neuroimaging initiative
AE	autoencoder
AG-CNN	attention guided convolution neural networks
AI	artificial intelligence
AIC	Akaike's information criterion
ALU	arithmetic logic unit
ANN	artificial neural network
AP	Antero-posterior
ASM	active shape model
ASPP	atrous spatial pyramid pooling
A-T	ataxia-telangiectasia
BCDCNN	breast cancer detection using CNN
BGRU	bidirectional gated recurrent unit
BN	batch normalization
BreCaHAD	breast cancer histopathological annotation and diagnosis dataset
BRIEF	binary robust independent elementary features
CAD	computer-aided design
CAD	computer-aided diagnosis
CAE	convolutional autoencoders
CCNN	cloud-based convolution neural network
CCTA	cardiac computed tomography angiography
CE	capsule endoscopy
CECT	contrast-enhanced computed tomography
CN	cognitive normal
CNN	convolutional neural network
CNNI-BCC	CNN improvement for breast cancer classification
CPH	Cox proportional hazard
CRF	conditional random field
CT	computed tomography

CVDL	cooperative variational profound learning
CVDs	cardiovascular diseases
CXR	chest x-ray
DAN	deep autoencoders
DaT	dopamine transporters
DB	database
DBN	deep belief networks
DBS	deep brain stimulation
DCT	discrete cosine transform
DG	disease group
DL	deep learning
DLA	deep learning analyzer
DNN	deep neural networks
DOG	difference of Gaussian
DR	diabetic retinopathy
DTI	diffusion tensor imaging
DTI	dynamic thermal imaging
DV	dorso-ventral
DWT	discrete wavelet transforms
ECG	electrocardiogram
EHR	electronic health record
EMD	experimental mode decay
FAST	features from accelerated segment test
FC Layer	fully connected layer
FDG	fluorodeoxyglucose
FFDM	full-field digital mammography
FFT	fast Fourier transform
FLANN	fast library for approximate nearest neighbors
FLOPs	floating-point operation
FNN	forward neural network
GANs	generative adversarial networks
GDWDL	greedy deep weighted dictionary learning model
GL	generative learning
GM	grey matter
GMM	Gaussian Markov model
GPi	globus pallidus
GRU	gated recurrent unit
HAM	human against machine
HCA	hierarchical clustering algorithm

HCI	human-computer interaction
HG	health group
HPF	high-power magnetic field
HRCT	high-resolution computed tomography
ICSS	intracranial self-activation
ILD	interstitial lung disease
KDD	knowledge discovery/data mining
KNN	K-nearest neighbor
LESH	local energy-based shape histogram
LR	learning rate
LRN	local response normalization
LSH	locality-sensitive hashing
LSTM	long short-term memory
LTP	local ternary pattern
MCCNN	mammogram classification using CNN
MCI	mild cognitive impairment
MIAS	Mammographic Image Analysis Society
ML	medio-lateral
MLP	multi-layer perceptron
MLPNN	multi-layer perceptron neural network
MMSE	mini-mental state examination
MNIST	Modified National Institute of Standards and Technology
MPSD	multiple patches in a single decision
MSOP	multi scale-oriented patches
NORB	NYU object recognition benchmark
OA	osteoarthritis
OCD	obsessive-compulsive disorder
PBC	patch-based classifier
PCA	principal component analysis
PD	Parkinson disorder
PET	positron emission tomography
PMF	probabilistic MF
PSF	point spread function
PSO	particle swarm optimization
RANSAC	random sample consensus
RBM	restricted Boltzmann machine
ReLU	rectified linear unit
ResNet	residual network
RNN	recurrent neural system

ROC	receiver operating characteristic
RPN	region proposal network
RTPCR	reverse transcription-polymerase chain reaction
SARS-CoV-2	syndrome coronavirus 2
SGD	stochastic gradient descent
SGRU	stacked bidirectional gated recurrent unit
SIFT	scale-invariant feature transform
SMOTE	synthetic minority oversampling technique
SNN	shallow neural networks
SN-VTA	substantia Nigra-ventral tegmental areas
SPSD	single patch in a single decision
SS	self-stimulation
STN	subthalamic nucleus
SURF	speed up robust feature
SVD	Saarbrucken voice database
SVM	support vector machine
SWS	Sturge-Weber syndrome
TB	tuberculosis
TDM	topological data mining
TDSN	tensor deep stacking network
TDV	total dermatoscopic values
UV	ultraviolet
VAE	variational autoencoder
VDM	visual data mining
VHL	von Hippel-Lindau
WCE	wireless capsule endoscopy
WHO	World Health Organization
WM	white matter

ACKNOWLEDGMENTS

This edited book's successful completion is made possible with the help and guidance received from various quarters. I would like to avail myself this opportunity to express my sincere thanks and gratitude to all of them.

I am deeply indebted to Almighty God for giving us this opportunity. I extend my deep sense of gratitude to our son, Master H. Jubin, and our parents for their moral support and encouragement at all stages of this book.

I extend my deep sense of gratitude to my scholars and friends for writing the chapters in time to complete this book. I sincerely thank my friends and family member for providing necessary prayer support.

Finally, I would like to take this opportunity to especially thank the Apple Academic Press for their kind help, encouragement, and moral support.

—*Editors*
E. Golden Julie, PhD
Y. Harold Robinson, PhD

I express my sincere thanks to the management of Vellore Institute of Technology, Vellore, India, and NVIDIA for providing a GPU granting support of her research work. Also, I would like to thank the Apple Academic Press for giving me the opportunity to edit this book.

—*S. M. Jaisakthi, PhD*

PREFACE

Deep learning (DL) is the most attractive technique in the field of biomedical engineering and health informatics during recent years because it provides an accurate diagnosis of disease. Accurate diagnoses of disease depend on image acquisition and interpretation. There are many methods to get high-resolution radiological images, but we still lack automated image interpretation. Currently, deep learning techniques are providing a good solution for the automatic diagnosis of disease with good accuracy. This book provides a clear understanding of how deep learning architecture can be applied in medical image analysis for providing automatic interpretations such as image segmentation, classification, registration, and computer-aided analysis in a wide variety of areas.

Chapter 1 reviews a variety of methods and techniques in the healthcare system using deep learning. The authors have used a number of deep learning tools such as K-nearest neighbor (KNN) algorithm, AlexNet, VGG-16, GoogLeNet, and vice versa for implementation. This chapter focuses on breast cancer analysis, lung tumor differentiation, pathology detection, and patient-learning using numerous methods.

Chapter 2 gives an overview of convolutional neural network architectures and their variants in medical diagnostics of cancer and COVID-19. The convolutional network is the basic architecture: based on these different variants like YOLO (you look only once), Faster RCNN, RCNN, AlexNet, and GoogLeNet are developed. Fast RCNN is based on region proposal network (RPN) for medical imagining. CNN variants have the ability to classify the images and detect the risk of diseases.

Chapter 3 discusses the technical assessment of various image stitching techniques in a deep learning approach. This chapter presents various image-stitching techniques based on different feature extraction methods. Image stitching can be regarded as a process of assembling more than one image of the same scene having an overlapping area in between them to make them into a single high-resolution image. The experimental results have revealed that ORB outperforms other methods in terms of rotation and scale invariance and execution time.

In Chapter 4, a deep learning approach for an acute neurocutaneous syndrome via cloud-based MRI images is discussed. A decision tree classification is an added advantage in CNN, which gives solutions for many different types of symptoms other than the MRI images. A set of pre-trained GoogLeNet libraries is used for the analysis of MRI images for this work. This chapter provides a more innovative cloud convolutional neural network for neurocutaneous syndrome in the biomedical field, which is under serious research.

Chapter 5, titled "Critical Investigation and Prototype Study on Deep Brain Stimulations: An Application of Biomedical Engineering in Healthcare," focuses on stimulating brain activities to the extent that is considered desirable to boost performance and be a productive human resource. This will be a promising product that can be viewed as a cure for various diseases closely related to brain activity. This includes conditions like epilepsy, Parkinson's disease, chronic pain, Dystonia, and even for normal people.

Chapter 6 gives an insight into various algorithms for medical image analysis using convolutional neural networks (deep learning). This chapter is aimed at the early prediction of Alzheimer's disease (AD). An algorithm that discriminates the mild cognitive impairment (MCI) and cognitive normal (CN) is casted-off, which shows better results in its analysis.

Chapter 7 deals with exploration of deep RNN architectures: LSTM and GRU in medical diagnostics of cardiovascular and neuro diseases. This chapter is delivers specific indispensable material about RNN-based deep learning and its solicitations in the pitch of biomedical engineering. This chapter will inspire young scientists and experts pioneering in the biomedical domain to swiftly comprehend the best-performing methods.

Chapter 8, titled "Medical Image Classification and Manifold Disease Identification through Convolutional Neural Networks: A Research Perspective," discusses a comprehensive analysis of various medical image classification approaches using convolutional neural networks (CNN). Here a short-term explanation of numerous datasets of medical images along with the approaches for facilitating the major diseases with CNN is discussed. All current progress in the image classification using CNN is analyzed and discoursed.

Chapter 9 discusses melanoma detection on skin lesion images using K-means algorithm and SVM classifier for detecting skin cancer in earlier stages. The proposed algorithm includes Sobel's process, Otsu's method, ABCD, and K-means with SVM classifier for getting the accuracy at 92%.

Chapter 10, titled "Role of Deep Learning Techniques in Detecting Skin Cancer: A Review," deals with automatic detection of skin cancer in the dermoscopic image that are required for detecting melanoma at an early stage. It deals with state-of-the-art deep learning from the foundation of machine learning as it provides better accuracy for medical images.

Chapter 11 explains deep learning and its applications in biomedical image processing. In this chapter, various blocks of DL are clearly discussed. Currently, DBN has evolved and proved to be the best when compared to the other networks. It may be noted that a particular network's accuracy entirely depends on the type of application and the features.

CHAPTER 1

REVIEW OF EXISTING SYSTEMS IN BIOMEDICAL USING DEEP LEARNING ALGORITHMS

M. SOWMIYA,[1] C. THILAGAVATHI,[1] M. RAJESWARI,[2] R. DIVYA,[3] K. ANUSREE,[3] and SHYAM KRISHNA[3]

[1]Assistant Professor of IT Department, M. Kumarasamy College of Engineering, Karur, Tamil Nadu, India, E-mails: sowmiyam.it@mkce.ac.in (M. Sowmiya), thilagavathic.it@mkce.ac.in (C. Thilagavathi)

[2]Professor in CSE, Sahrdaya College of Engineering and Technology, Kerala, India, E-mail: rajeswarim@sahrdaya.ac.in

[3]Assistant Professor in CSE, Sahrdaya College of Engineering and Technology, Kerala, India, E-mails: divyar@sahrdaya.ac.in (R. Divya), anusreek@sahrdaya.ac.in (K. Anusree), shyamkrishna@sahrdaya.ac.in (S. Krishna)

ABSTRACT

In recent days, many researchers focus on interdisciplinary research work to solve various research problems. One such interdisciplinary research work is the implementation of deep learning (DL) in biomedical engineering and health informatics. It is mandatory to review various research works by various authors, which are implemented using DL in the above fields in order to understand the problem domain, different DL methods/techniques/theorems for prediction and analysis. The main intention of this chapter is to review a variety of methods and techniques in the healthcare system using DL. Authors have used a number of DL tools such as K-nearest neighbor (KNN) algorithm, AlexNet, VGG-16, GoogLeNet, and vice versa for implementation. It has been noticed that DL architecture has been formulated for medical analysis, such as various cancer prediction, in order to predict the accuracy of results. This DL architecture has three basic layers that help to train and test the model, which are the input layer, multiple hidden layers, and the output layer. The output of various medical analyzes depends on the DL architecture that is used along with a number of convolutional hidden layers. In this chapter, breast cancer analysis, lung tumor differentiation, pathology detection, patient-learning using numerous methods such as similarity learning, predictive similarity learning, and adaptive learning using the DL approach has been described.

1.1   INTRODUCTION

The medicinal industry multiplies because of the headway of remote correspondence innovation and AI applications. The quantity of maturing individuals in the world advances, though the proportion of particular specialists to patients diminishes. Individuals become incredibly involved, and traffic blockage increases. Late advances in remote correspondence innovation have empowered the improvement of a savvy human service structure that is quick, consistent, and universal. The most recent improvements in AI, for example, deep learning (DL), enhancing the exactness of frameworks before utilizing an extensive measure of information. Remote correspondence innovation and AI calculations can understand this issue to a limited degree. Hence, much research is led to build up a keen human services system utilizing 5G innovation, edge, and distributed computing, and DL.

   In the most recent decennary, DL algorithms like recurrent neural system (RNN), convolutional neural system (CNN), and autoencoder (AE) was the most commonly used type of DL models, for example, mechanical structure [20] and picture acknowledgment [21]. Recently, a few works have connected variational autoencoder (VAE) [22] to perform CF task in suggestion, for example, CVAE [23], CAVAE [24], CLVAE [25], and VAECF [26]. VAE has a vulnerability in recommender systems with huge information and the capacity of catching non-linearity, and it is a non-direct probabilistic model. In spite of the adequacy of these VAE-based strategies, there are yet a few disadvantages; for example, to separate the idle vectors, CVAE and CAVAE legitimately employ content data. In essential consideration frameworks, the extra data of patients and specialists are exceptionally well-off, and was not completely used for the betterment of proposal execution, which creates HRS even in their earliest stages concerning dependability and unwavering quality. To take care of those issues above, a CVDL (cooperative variational profound learning) model is devised for HRS in essential consideration, to give personalization into the patient's consideration and to give the understanding by utilizing the patient's inclinations.

CVDL makes both inert client/thing vectors through a different neural system structure, which can viably study things for further CF process and non-direct idle portrayals of clients. EHR (electronic health record) frameworks store information related with every patient experience, including statistical data, analyze, lab tests and results, remedies, radiological pictures, clinical notes, and more [1]. While fundamentally intended for improving human services effectiveness from an operational point of view, numerous investigations have discovered auxiliary use for clinical informatics applications [5, 6]. In this chapter, the particular DL system was utilized for EHR information examination and deduction, and talked about the solid clinical applications empowered by these advances. Dissimilar to other ongoing overviews which audit DL in the expansive setting of wellbeing informatics applications going from genomic examination to biomedical picture investigation, this review is centered solely around profound learning procedures custom-fitted to EHR information. In opposition to the choice of particular, viable applications found in these reviews, EHR-based issue settings are described by the heterogeneity and structure of their information sources and by the assortment of their applications. Tolerant closeness learning is a basic and significant assignment in the human services area, which improves clinical basic leadership without bringing about extra endeavors from doctors. The objective of patient closeness is to get familiar with the important metric unit, which estimates the relational similitudes among patient sets as per patient's wellbeing files.

A legitimate closeness value empowers different downstream applications, for example, customized prescription [1, 2], medicinal conclusions [3], direction investigation [4], and accomplice study. The pervasiveness and developing electronic volume wellbeing records (EHRs) gives phenomenal chances to get better medical choice help. The EHR information, which provides the patient's longitudinal electronic record, is a profitable hotspot for prescient demonstrating, which will help the healthcare industry. The records of wellbeings are transiently ordered using victim encounters spoke to as many advanced clinical occasions (for example, restorative codes). Mining EHRs is particularly testing contrasted with standard information mining assignments, because of its loud, unpredictable, and heterogeneous nature for the wellbeing's record information, the computation of physically affected person visits differs to a great extent, because of patients' unpredictable visits and inadequate chronicles. The previously mentioned learning measurements can't be straightforwardly connected to the longitudinal information, as the chronicled files of every patient don't normally frame a practically identical vector. Accordingly, one of the key difficulties in estimating persistent likeness is to determine a successful portrayal for every patient without the loss of his/her authentic data.

The first depends on triplet misfortune work, which learns an edge to isolate the separation of negative and positive examples. Thusly, it is good to get a separation worth demonstrating the general comparability between patients. Secondly, is to perform characterization over scholarly portrayals with the plus mark for comparative combines and minus mark for different sets. In addition, the closeness likelihood among couples of patients demonstrates the level of risk between two patients building up a similar illness; it is used as the point to evaluate the likeness over patients. Subsequent to acquiring the closeness data, two undertakings are performed: ailment forecast and patient grouping, which are application regions of customized human services, so as to approve the educated measurements.

1.2 BREAST CANCER ANALYSIS USING DEEP LEARNING (DL)

Zhang et al. [13] stated the concept of selection of features and separation models in DL, which is used for predicting breast cancer. The flowchart for unsupervised and supervised learning methods for breast cancer was demonstrated by the authors. Authors implemented the concept, which includes data alignment, the objective function for feature extraction that consists of nonlinear transformation and reconstruction loss, activation function in auto-encoder, optimization with ADAM, normalized initialization, and Adaboost algorithm for classifier learning.

Sellami et al. [28] elaborated sequence exploration on breast cancer ultrasound pictures by using BI-RADS characteristic extraction. The authors discussed the preprocessing and segmentation process, which includes digital image processing for spot removal and picture segmentation. Morphological features were classified according to the BI-RADS lexicon, which includes three classes such as the pattern, pattern position, and orientation. Texture features were extracted using three classes of the lexicon, such as boundary classification of the lesion, classification of echo pattern, and posterior acoustic feature classification of logical methods of posterior acoustic characteristic.

Gubern-Merida et al. [31] defined a completely automatic framework for breast partition and density estimation. The authors provided a general overview of dense tissue segmentation. There are three preprocessing algorithms applied, such as the N3 bias field correction algorithm, detection of the sternum, and normalization of intensities of MR images. Breast body segmentation has been elaborated. Breast density segmentation has been carried out and evaluated as programed bosom division and thickness division utilizing EM.

Li et al. [32] stated the concept of selective element mining for breast cancer for which histopathology picture categorization was performed using completely convolutional AE. Breast cancer biopsy image dataset was collected to create the likelihood guide of harmful cells. Authors used a patch-based learning solution, which is depicted in Figure 1.1. First-order statistics of true-normal patches were given as the input during the training phase using one-class SVM (support vector machine) for malignant patch detection. SVM's mapping functions were given as the input for training phase 3, which uses the 1-layer neural system for Platt's score and produced patch posterior probability as output.

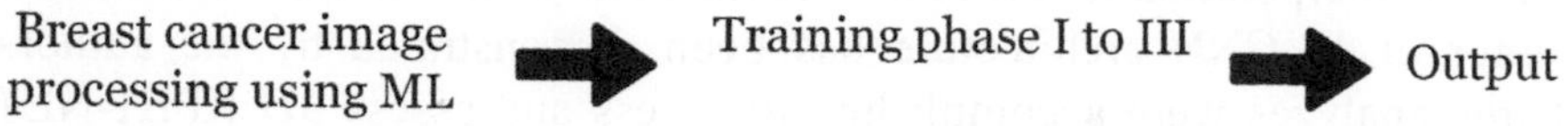

Breast cancer image processing using ML ➡ Training phase I to III ➡ Output

FIGURE 1.1 Selective element mining for breast cancer.

Athreya et al. [33] used the concept of ML, which guides distinguish drug mechanism in breast cancer. Bosom malignancy tissue comprising cancer cells was done using a human genome of 24,000 genes. Gene expression matrix was constructed using 192 cells of 24,000 × 192. Metformin has been applied to ensure whether it affects gene expression of few cells or many cells. Machine learning approaches were introduced to identify a list of candidate genes using unsupervised learning methods and pathway analysis. From unsupervised learning, it is noticed that cells, which are exposed to the drug, may tend to represent differences in their gene expression that are found in molecular interactions. Data characteristics and preprocessing were performed, which states that 80% of the qualities were latent in the information while preprocessing in which only 5% of genes have shown with changes using metformin.

1.3   LIVER TUMOR ANALYSIS USING DEEP LEARNING (DL)

Trivizakis et al. [34] has extended the DL concept for liver tumor differentiation. MRI is treated to be a powerful tool for detecting small lesions that leads to malignancy in the liver tumor. The authors depicted a 3D convolutional neural network (CNN). Data augmentation is also considered to be important in original patch analysis using image deformation, 270° rotation, and so on. Tissue classification was performed using SVM with various evaluation parameters such as accuracy, sensitivity or recall, and precision, an overall analysis of 2D and 3D.

1.4   DEVELOP 3D PET FOR RADIONICS EXAMINATION AND OUTCOME PROGNOSIS

Amyar et al. [41] have collected data of 97 patients having esophageal cancer for which they have applied PET with CT during the initial stage. Two 3D convolutional layers were used to define 3D RPET-NET, for which image preprocessing was done on the dataset using the K-nearest neighbor (KNN) interpolation algorithm. Visualization of a 2-D cut of a divided tumor of 1S-CNN architecture has been demonstrated by the authors. Three analyzes were accomplished to assess and assess 3D RPET-NET. The outcomes were contrasted with 3 RF-based methodologies. Cross-validation has been performed by splitting the entire data into two groups, which will be used for training and testing. The first group was intended for training the models with 77 patients, and the second group was used for testing the model with 20 patients. Alternatively, training samples, such as 77 patients, were divided into two groups. Fifty-five patients details are used to train the set, and the remaining 22 patients are used for the validation set.

1.5   A GREEDY DEEP LEARNING (DL) METHOD FOR MEDICAL DIAGNOSIS

Greedy deep, weighted dictionary learning [42] which is used for medical disease analysis to overcome over-fitting problem while classifying and training patient data. Internet of Things, together with the healthcare system, has been utilized to enhance the dependability in the analysis of portable media, which is intended for medicinal services to anticipate the sicknesses and for medical diagnosis. ADHD-200 (attention deficit hyperactivity disorder) database (DB) were used to extract 30 datasets of depression as a training set. Authors have defined a Boltzmann machine with three hidden layers which involves unsupervised learning model to carry out the processing along with feedback mechanism. Dictionary learning was done using sparse coding from matrix decomposition.

Greedy deep weighted dictionary learning model (GDWDL) was proposed by the authors for performing clinical data preprocessing, which is followed by extracting data information was done using type series analysis. Two groups were formed by the authors such as health group (HG) and the disease group (DG) in order to segregate the patient details. The optimization process was carried out in the single-layer neural network by using shallow dictionary learning, which is said to be a non-convex optimization method. The algorithm was provided by the authors to find out the objective function to achieve the optimal solution. Comparison has been made by the authors with different dictionary size for various sensitivity of algorithms such as FDDL, DFDDL, and GDWDL.

1.6 MULTI-ARRANGE PROFOUND LEARNING FOR MEDICAL DIAGNOSIS

Yan et al. [43] stated the concept on multi-instance DL, which is used to discover discriminative local anatomies in medical diagnosis. Body part identification was indicated by local image information. The authors designed a DL framework which contains multiple stages in order to recognize body parts of a patient by applying image classification. CNN was used to determine the local regions of a human body that are sensitive. Authors applied cut-based body part identification, which is usually a multi-class picture arrangement technique which includes four kinds of geometry components, for example, square, circle, triangle, and diamond are utilized. Classification accuracy with respect to various classes for triangle and square and then diamond and circle as far as review, accuracy, and F1 score in percentage were tabulated by the authors.

1.7 SKIN LESION GROUPING UTILIZING PROFOUND LEARNING

Marwan [44] used a convolutional neural system with a novel regularizer for skin injury arrangement. The authors used CNN along with the regularization technique in order to control the complexity of the classifier. Regularization process involves assigning weight matrics along with the regularization parameter, which will work in a convolution filter. In addition, 5600 images were used for training, and 2400 images were used for validation. Out of which 4533 were malignant and 19,373 benign skin lesion images were found out.

1.8 STAIN-INVARIANT CONVOLUTIONAL NETWORKS

David et al. [45] trained refined stain-invariant arrangement for entire slide mitosis discovery for which PHH3 was utilized as a source of perspective in H&E bosom histology. Mitotic activity utilizing PHH3 stained slides consists of a whole-slide image, which is used for candidate detection using blue and brown channels. Then PHH3 whole-slide images with candidates were used for sampling and manual annotation, which will be given for training using CNN. H&E Mark: Preparing a Mitosis indicator includes assembling a training data set; stain augmentation, which consists of three invariances such as morphology, stain, and artifact invariance, then ensemble and network distillation, which is followed by the outcome at the whole-slide level.

1.9 HCI-KDD APPROACH

The work proposes an HCI-KDD approach [46], which combines various methodological approaches of DL for the healthcare sector. The approach is a mere combination of HCI (human-computer interaction) and KDD (knowledge discovery/data mining). The former emphasis on cognitive science and the later on machine learning. HCI focuses on specific experimental methodologies and KDD on computational learning problems.

The main task in HI is the data ecosystem identification. Mainly four types of data pools are made based on the context of data origin. The four data pools are: biomedical research data, clinical data, health business data, and primitive patient data. Data preprocessing step is followed by a data integration step. The information integration can overcome the drawbacks laid among medical and biological research points and issues.

The information reconciliation procedure joins information from different sources and gives a brought together view about the information. Similar concept is data interpretation, which matches various data which points one object into a single consistent representation. It is found that fused data is having much better information than the original large collection of data.

Current trend of machine learning is towards the automated machine learning (aML) algorithms, which completely expels the human interventions so that the process becomes completely automated. Voice identification and processing, recommendation systems, automated vehicles are examples of the above mentioned. Another area of ML is the interactive machine learning (iML), which can associate with both statistical assistants and human assistants and they have optimized experiencing via such methods of collaborations. However, the iML algorithms face certain issues-much more difficult, time consuming, hard to replicate, and robustness.

Discovering the associations among data items and to map the necessary data structures, graph theory can be used. Graph theoretical algorithms help us to map the concepts of computer networks, network analysis, and data mining and cluster analysis to the HI concepts. Complexity occurring in the graphical representation of data makes it more difficult to go with Graph theory algorithms. Topological data mining methods (TDM) are similar and related to graph theory-based methods. Homology and persistence are the popular techniques of TDM. The cycles of each space determines its connectivity. The groups which are formed from such sequences are called homology aggregations, which can be computed with the help of linear algebra using an explicit description of the space. A notable measure that could be utilized to quantity the proximity is Cosine uniformity measure. Entropy of an information accounts for the uncertainty which persists in the data. In the graph theory concept, graph entropy is described to measure the structural information of the graph data.

Another important concern in the area of study is the visualization of the data. Visual data mining (VDM) bolsters intuitive and adaptable system representation and the examination of information. Clustering helps us in data visualization-it recognizes homogeneous gatherings of information dependent on the provided measurements.

1.10 PROFOUND NEURAL MODELS FOR FORECAST IN HUMAN SERVICES

This work [47] proposes deep neural architectures that can accept raw data as input and can provide desired outputs. The work primarily concentrates on neurodegenerative maladies-Parkinson's infection. The system is to support nurses and doctors to provide advanced and authentic prognostic and analysis about the disease.

They created public data set with 55 patients having Parkinson's disease and 23 with the interrelated complaints. MRI scan data can show the extent to which the brain has degenerated. So the MRI data is analyzed to identify the lentiform nucleus and capita of caudate nucleus. The image sequences are processed in batches to get the volume information of these interest points.

Dopamine transporters (DaT) scan is another imaging technique used to populate the DB. The degree of dopaminergic innervations to Striatum from Substantia Nigra can be easily identified. From the series of images available via DaT scan, the doctor identifies the most representative one. He then marks the areas corresponding to the most representative scan image. These corresponding spots are compared with the neutral points by using an automated system and produce the ratios which can be used for comparing purposes.

In addition to MRI scan data and DaT scan data, clinical data of a patient is taken to get an idea about the motor/non-motor experiences, motor examinations, and complications of daily experience of the subject. For the training purpose of the proposed model, an annotated representation of MRI data and DaT data is considered. They proposed a deep neural architecture for the diagnosis and prediction of the disease. The components of the proposed model are: Deep CNN, Transfer learning, and recurrent neural networks.

The deep CNN exploits the spatial information of the input. The learning failures which occur due to over fitting when training is done with complex CNN's with small data can be avoided to an extent by using the concept of transfer learning. The concept of transfer learning is that, we will use the previously used networks which are trained with bulky image datasets which are fine-tuned as per our needs or part of the network for the training purpose using the tiny dataset. Sequential data can be processed with the help of a recurrent neural network (RNN). Experiments show that among the various RNNs, gated recurrent units (GRUs) have better performance, and it would be utilized for the prediction and assessment of Parkinson's disorder. CNN's consist of enormous internal representation of the input data. The

CNN under study have 50 layers. In addition to these layers, there are three numbers of fully connected layers (FC layer)-rectified linear units (ReLU).

The MRI and DaT scan data are given as input to the network. If our area of interest is with epidemiological and clinical data inputs, those inputs are given directly to the FC1 layer. In the CNN, the linear FC3 layer performs the estimation of clinical data. The input to the RNN part is F1, F2, F3, …, FN, and the corresponding expected outputs are O(1), …, O(N). A single input to the architecture is a set of four images out of which three consecutive images are grayscale from the MRI data and one color image from DaT scan.

The implementation of the architecture starts with the transfer learning of weights of the ResNet network. Tensor Flow platform is the tool which they have used for implementing the software part of the architecture. The architecture can be enhanced for prediction and assessment of degenerative diseases.

1.11 PATHOLOGY DETECTION USING DEEP LEARNING (DL)

Voice pathology [37] can be extreme if there should arise an occurrence of disappointment in early identification and appropriate administration. This pathology is especially predominant among experts, for example, instructors, vocalists, and legal advisors, who too much utilize their voice. In this chapter, another versatile human services structure has been elaborated that contains a programmed voice pathology recognition framework. The system comprises of brilliant sensors, distributed computing, and correspondence among patients and partners, for example, emergency clinics, specialists, and medical caretakers. The proposed voice pathology recognition framework utilizes convolutional neural system (CNN). A few tests were performed utilizing an openly accessible DB known as SVD (Saarbrucken voice database). A few frameworks utilize acoustic highlights [6], for example, Mel recurrence cepstral coefficients [14], coefficients of perceptual direct forecast, and straight prescient cepstral. The highlights are embedded from speaker and discourse acknowledgment applications which are additionally utilized unitedly with understood classifiers, for example, concealed Markov systems, Gaussian blend system, bolster vector system [7], and counterfeit neural system [10].

Tang et al. [35] stated the concept for survival analysis using DL CNN. There are three stages in data preprocessing, which consists of inspecting from WSIs, characteristic extrication, and bunching then finally choosing groups. This structure includes the image size of 128×128, which is given as the input for a convolutional layer that performs dynamic routing. A survival cap, which includes long term and short term, produces the output, which is linked with survival loss that includes margin loss, cox loss, and reconstruction loss.

Amin et al. [7] explained the concept of a cognitive smart healthcare system, which is for pathology detection and monitoring. The authors described the cognitive smart healthcare system scenario, which includes IoT smart sensors, cognitive engine and DL server and other interconnecting devices, which is depicted in Figure 1.2.

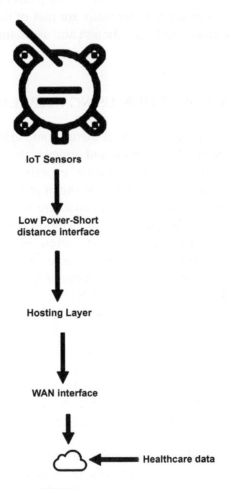

FIGURE 1.2   Cognitive smart healthcare system.

Different EEG preprocessing and representation techniques were used for pathology detection and classification. The authors used DL tools such as VGG-16 and AlexNet to simulate the results.

Alhussein and Muhammad [59] elaborated voice pathology identification utilizing profound learning on portable human services systems. Exemplification of the same is given in Figure 1.3, which includes the required mobile healthcare system and the interaction among them. This mobile healthcare framework and the corresponding voice signal processing used for pathology detection.

Input sensors with local station

Authentication

Processing
Cloud Manager

Registered doctors and nurses

FIGURE 1.3   The mobile healthcare framework.

The authors also depicted the architecture of the VGG 16 DL network and the architecture of CaffeNet for processing the voice signal.

Voice pathology might be surveyed in two different methods: subjective and objective. In subjective evaluation, a professional physician assesses voice pathology through hearing or through the use of tools, including a stroboscope or laryngoscope. In positive activities, more than one medical doctor examines a similar case to accomplish an accord. As anticipated, a particular wellbeing specialist will settle on the last decision with respect to voice pathology [8]. Goal evaluation is directed through a PC method, which examines the voice example and settles on a preference dependent on the assessment outcomes. This method of evaluation is fair-minded against subjects. In objective evaluation, the influenced individual transfers his voice sign to an enlisted human services structure in the wake of paying the enrolment charge. In this chapter, it is extraordinarily engaged in the concerned appraisal of voice pathology identification. To this point, various writing offers many voice pathology identification structures. The early frameworks utilized voice parameters, including gleam, jitter, sign-to-noise proportion, commotion symphonious proportion, and glottal-to-clamor proportion [11]. Parameters are measurements of voice top notch. A solitary parameter demonstrates deficient to identify pathology consequently, in a few structures; numerous parameters are mixed to harvest higher results [12].

Numerous systems use acoustic features [6] inclusive of Mel frequency cepstral coefficients [14], direct prescient cepstral coefficients, and perceptual straight expectation coefficients. Those highlights are imported from discourse and speaker acknowledgment bundles. These highlights likewise are utilized aggregately with a popular classifier, which incorporates concealed Markov models, Gaussian blend model, guide vector machine [10], and synthetic neural community [10].

Recently, deep getting to know primarily voice pathology recognition frameworks carried out with enhancing the accuracy [17, 19]. Specific models of deep studying [18] had been determined from picture preparing packages. The structures, as a rule, transform time-space voice signals into spectrograms that might be considered as photos. The fashions that have been utilized in voice pathology exploration consist of VGG-sixteen [20], AlexNet, and CaffeNet. Those models are pertained using tens of millions of snapshots. However, those deep gaining knowledge of-primarily based structures reject pathology characterization. These days, voice pathology discovery structures are incorporated into a social insurance system which contains the utilization of huge certainties. Various structures use edge processing to offload the transfer speed prerequisite [30]. Nothing unless there are other options noted structures utilizes parallel CNNs to abuse the time-segmental components of voice signals. Twenty-four melanoma sufferers have been enrolled in the observation and CRP degrees have been received.

The CRP estimations [8] were done on weekdays, with certain exemptions where everyday estimations have been made. The utilization of a rotator, plasma progressed toward becoming remote from the total blood assembled, subsequent to evacuating portable and protein flotsam and jetsam; aliquot and put away at 80°C for latter implementation. The stages were chosen through a research facility chemical associated immunosorbent measure (ELISA). The examination is a plate-basically dependent which examines strategy intended for identifying and evaluating peptides, proteins, antibodies, hormones. Sooner than tutoring, a succession of preprocessing tasks are executed, for example, insights institutionalization to increase 0 mean and unit change, and sign destroying utilizing experimental mode decay (EMD). After the investigations, the forecasts are unstandardized the utilization of the parameters determined ahead of time. The RMSE-root-imply-rectangular botches is determined from the unstableness's expectations that needs analyzing the conjecture values with real CRP perceptions, RMSE = $\sqrt{1/N\sum_{i=0}^{N}(x-xi)^2}$ where xi is anticipated value, Oxi speaks to observed value, and N shows the quantity of tests. The RMSE is generally utilized in various studies which includes regression technique. Reality that RMSE punishes a greater divergence from the mean brings us to lease this measurement inside the investigation. CTR employs item content material to improve CF methods and has accomplished encouraging execution by coordinating both individual score and article content. CTR consolidates the benefits of each probabilistic MF (PMF) and theme displaying (LDA) designs, and comprises of the idle variable for balancing the subject extents while demonstrating the client rankings and the balance variable can successfully catch the thing inclination for a specific consumer thinking about their rankings.

CTR doesn't make the most client records and can't look at reliable idle client portrayals. To address this issue, a couple of examines has been introduced utilizing extraordinary versions which accommodates general records into CTR. In addition, CTRSMF [15] and C-CTR-SMF2 [16] included CTR with SMF version the use of a method this is just like SoRec, wherein the social connections are simultaneously factorized with the score lattice. Be that as it may, they don't screen the fundamental relatives among clients because of the deficiency of physical legitimization. In contrast with CTRSMF and C-CTR-SMF2, LACTR, and RCTR legitimately inspect the portion of premium when clients apportion to different clients and use this scholarly impact to soothe spared issue. The strategies depend on that the social cooperation's of clients generally agree to topically practically identical substance, so they might be sensitive to stand-out types of datasets, and the forecast precision may likewise fluctuate with the circulations of datasets. For social suggestion, CTRSTE coordinates client rankings, thing substance and consider troupe into CTR, which is straightforward in the algorithmic statute. However, its portrayal usefulness is obliged on account of LDA model, and the dormant portrayal educated isn't constantly viable enough while the side certainties is scanty.

With the aid of assessment, CTRDAE utilizes DAE and LDA to shopper general portrayal and data exemplification individually, to spare you buyer relationship overfitting under the inadequate social individuals from the family circumstances. However, the content illustration ability of CTRDAE is equivalent to CTR, that's restricted because of subject matter modeling version. Although those works have stepped forward CTR in separate components by the use of either substance or informal community insights, an essential issue stays, i.e., how to correctly integrate article substance, client evaluations and person profiles/family members into CTR [29]. Unlike preceding CTR-primarily based recommendation techniques, this concept develops the propagative techniques of clients and devices using a neural erratic structure that allows this method to catch non-straight inactive portrayals of the two clients and items.

1.12   LEARNING MODELS IN DEEP LEARNING (DL)

1.12.1   SIMILARITY LEARNING

For a new person, figuring out historic facts of sufferers who're identical might support recover comparable recommendations for foreseeing the logical results of the newly affected person. In 2017 [1], mixed affected person comparability and medication closeness evaluation and introduced a contrasting brand proliferation technique to pick out the correct drug compelling for a distressed person. In exercise, distinctive doctors have distinctive realization of affected person similitude dependent on the points of interest. The use of doctor criticism as the supervision [9] introduced a regionally administered metric mastering (LSML) set of rules that attains a comprehensive Mahalanobis distance. For the reason that getting doctors' advice is intense and profoundly estimated in truth, Wang and Sun introduced a pitifully regulated patient similitude acing technique which best utilizes a little amount of supervision data given by the doctors.

1.12.2 PERSONALIZED HEALTHCARE [38]

As of late, the customized expectation in medicinal services projects gets a developing pastime from analysts. It plans to discover the remarkable qualities for individual sufferers, and perform focused on understanding specific forecasts, suggestions, and medications. The vast majority of the works implement customized prediction with the aid of matching scientific comparable sufferers. Authors completed a comparative take a look at worldwide, nearby, and customized modeling, and discovered that altered methods could receive higher overall performance across exceptional bioinformatics class duties.

The aforementioned techniques require the entry of every affected person as a vector. A conventional manner is to acquire Function vectors by means of the usage of the static facts of sufferers consisting of demographic, and facts records (e.g., sum, common, and so forth) within a positive time variety, as the patient illustration. However, these handmade characteristic vectors absolutely ignore the temporal family members throughout go to sequences. To represent the transient actualities, utilized a powerful programming set of guidelines to discover the surest close by arrangements of patient successions [27] advanced two answers for patient closeness contemplating, unaided, and regulated, the utilization of a CNN-based absolutely comparability coordinating structure; and established a second RNN for powerful fleeting coordinating of affected groupings to procure the similitude positioning.

In this segment, first, how to examine a powerful illustration for the longitudinal EHR facts was shown, introduced two techniques to degree the correspondence among pairs of patients. Using discovered equivalence statistics, two tasks were carried out for customized human services: infirmity expectation and patient grouping.

1.12.3 EXEMPLIFICATION LEARNING

1.12.3.1 FUNDAMENTAL ANNOTATIONS

Prosperity documentation of a patient consists of a progression of campaign details, the restorative canon are registered demonstrating the infection or therapy the patient endured or got. The canons can be delineated to the International Classification of Disease (ICD-9).1 It is suggested that all the stand-apart helpful canons from the EHR information as c1, c2. c|C| ∈C, where |C| is the measure of magnificent remedial canons. Expecting that there are N affected people, the n-th affected has various visits Tn. A patient pn can be spoken to by a grouping of visits represented as V1, V2, VTn. Each visit Vi is meant by a high dimensional twofold vector vi ∈ {0, 1}|C|, demonstrating whether Vi contains the canon cj or not.

1.12.4   SIMILARITY LEARNING

Contemplating the comparison between every pair of sufferers is the key advance for customized human services. There are two strategies to gauge the likeness among influenced individual vectors found from Section 1.12.3.1, SoftMax oriented absolutely structure and ternary misfortune system.

1.12.4.1   PREDICTIVE SIMILARITY LEARNING

The likeness among a couple of listing could be estimated with the guide of a bilinear separation: S = hi Mhj, wherein the coordinating lattice M ∈ Rm×m is symmetric. To make sure that the symmetric imperative of M, its miles decayed as M = LT L, where L ∈ Rl×m with l < m to make certain a low-position include.

A symmetric requirement was recalled for listing link and transfer influenced individual listing to avail a similitude vector to guarantee that the request for sufferer has no effect on the likeness score. Initially, convert hello there and hj into separate listing along with the estimation utilizing the framework.

From that point onward, H and S are linked and after that encouraged into a totally related SoftMax layer, to get a yield opportunity y' that takes the path of least resistance cost among 0 and 1. In addition, the floor reality y is set as 1 if two patients have the risk of building up the equivalent sickness, in some other case 0. The higher estimation of y' implies the higher likelihood that pi and pj contain position with a consistent category, or two patients have littler separation and are increasingly like one another. The model can be prepared from start to finish, and every one of the parameters are refreshed at the same time.

Triplet-Loss Metric Learning: Metric learning plans to become familiar with an appropriate separation metric for a specific errand, which is significant to the exhibition of numerous calculations. The possibility of metric learning is utilized to become familiar with the overall separation of sufferers. In the customary measurement method, a straight change L is utilized to delineate crude information into another dimension. Alternative measurement in that dimension can call more likely measure the overall separation of info occasions. The distance between instances x_i and x_j could be generated by applying the Euclidean distance in the alternative dimension.

In profound measurement method, the direct change L is supplanted by a neural system f to become familiar with the convoluted nonlinear connections between crude highlights. In this issue of patient similitude learning, this nonlinear change is found out through the CNN activity. Triplet loss [39] is used as the objective function which carries a hard and fast of triplets, in which each triplet has an anchor, a fantastic, and a poor instance. An effective pattern contains the identical magnificence content because the dependence with bad pattern may contain the distinctive class label.

The one-hot EHR lattice of patient pi is correlated to an implanting framework, and after that sustained into CNN to get a listing portrayal. pi+ and pi− distribute indistinguishable limits from pi. Pairwise separations could be determined depends on the listing portrayals, and ternary misfortune is utilized to refresh every one of the parameters, for example, the separation between stay pi and positive example p+i ought to be nearer than the separation among pi using certain arranged boundary. This measurement picking up learning of layer is expedited top of CNN that considers the listing delineation utilizing the contribution to compute separation among sufferers. The capacity is limited using returned engendering, and the majority of the parameters are forward-thinking at the same time. The educated separation metric demonstrates the comparability among influenced individual sets, with littler separation esteems for better closeness. The system of triplet-misfortune essentially based profound likeness acing is a conclusion to-stop becoming more acquainted with the system.

1.14   CONCLUSION

DL is assumed to be a subset of machine learning, which provides the architecture, algorithm, and methods that are inspired by the structure and function of the human brain. The main functionality of DL is that the designed system will learn and adapt by itself to new data and situation. DL requires a large volume of dataset for training and testing. Hence, DL is a suitable method to apply in most of the interdisciplinary areas such as biomedical engineering and health informatics since they produce huge volumes of data day by day in the healthcare system. In recent days, a huge volume of genetic information, RNA/DNA sequences, amino acid sequences, and other patient-related data are getting generated, which are to be processed effectively by using the DL approaches. This chapter provided a review of various DL techniques in the medical healthcare system for various applications, which include breast cancer analysis, lung tumor differentiation, voice pathology detection, current systems in e-healthcare. This review revealed that DL is found to be an effective method to deal with medical data to enhance performance.

KEYWORDS

- AlexNet
- attention deficit hyperactivity disorder
- cancer prediction
- convolutional neural system
- deep learning
- similarity learning

REFERENCES

1. Chen, M., Hao, Y., Hwang, K., Wang, L., & Wang, L., (2017). Disease prediction by machine learning over big data from healthcare communities. *IEEE Access, 5,* 8869–8879.
2. Al-Nasheri, A., et al., (2017). Voice pathology detection and classification using auto-correlation and entropy features in different frequency regions. *IEEE Access, 6,* 6961–6974.
3. Boyanov, B., & Hadjitodorov, S., (1997). Acoustic analysis of pathological voices: A voice analysis system for the screening of laryngeal diseases. *IEEE Eng. Med. Biol. Mag., 16*(4), 74–82.
4. Lopes, L. W., et al., (2017). Accuracy of acoustic analysis measurements in the evaluation of patients with different laryngeal diagnoses. *J. Voice, 31*(3), 382.e15–382.e26.
5. Jia, Y., et al., (2014). Caffe: Convolutional architecture for fast feature embedding. In: *Proc. 22nd ACMInt. Conf. Multimedia (MM)* (pp. 675–678).

6. Ali, Z., Muhammad, G., & Alhamid, M. F., (2017). An automatic health monitoring system for patients suffering from voice complications in smart cities. *IEEE Access*, *5*, 3900–3908.

7. Amin, S. U., et al., (2019). Cognitive smart healthcare for pathology detection and monitoring. *IEEE Access*, *7*, 10745–10753.

8. Hossain, M. S., & Muhammad, G., (2019). Emotion recognition using deep learning approach from audio-visual emotional big data. *Inf. Fusion*, *49*, 69–78.

9. Kalogirou, C., et al., (2017). Preoperative C-reactive protein values as a potential component in outcome prediction models of metastasized renal cell carcinoma patients receiving cytoreductive nephrectomy. *Urol. Int.*, *99*(3), 297–307. doi: 10.1159/000475932.

10. Adel, M., et al., (2016). Preoperative SCC antigen, CRP serum levels, and lymph node density in oral squamous cell carcinoma. *Med. (Baltimore)*, *95*, e3149. doi: 10.1097/MD.0000000000003149.

11. Steffens, S., et al., (2013). High CRP values predict poor survival in patients with penile cancer. *BMC Cancer*, *13*(1), 223. doi: 10.1186/1471-2407-13-223.

12. Coventry, B. J., Ashdown, M. L., Quinn, M. A., Markovic, S. N., Yatomi-Clarke, S. L., & Robinson, A. P., (2009). *CRP Identities Homeostatic Cimmune Oscillations in Cancer Patients: A Potential Treatment Targeting Tool?* (Vol. 7, p. 102). doi: 10.1186/1479-5876-7-102.

13. Dejun, Z., Lu, Z., Xionghui, Z., & Fazhi, H., (2018). *Integrating Feature Selection and Feature Extraction Methods with Deep Learning to Predict Clinical Outcome of Breast Cancer* (Vol. 6). doi: 10.1109/ACCESS.2018.2837654.

14. Blank, C. U., Haanen, J. B., Ribas, A., & Schumacher, T. N., (2016). The 'cancer immunogram.' *Science*, *352*, 658–660. doi: 10.1126/science.aaf2834.

15. Brustugun, O. T., Sprauten, M., & Helland, A., (2016). C-reactive protein (CRP) as a predictive marker for immunotherapy in lung cancer. *J. Clin. Oncol.*, *34*, Art. no. e20623. doi: 1 0.1200/JCO.2016.34.15_suppl.e20623.

16. Teishima, J., et al., (2015). The impact of change in serum C-reactive protein level on the prediction of effects of molecular targeted therapy in patients with metastatic renal cell carcinoma. *BJU Int.*, *117*(6)B, 67–74. doi: 10.1111/bju.13260.

17. Tomita, N., Cheung, Y. Y., & Hassanpour, S., (2018). Deep neural networks for automatic detection of osteoporotic vertebral fractures on CT scans. *Comput. Biol. Med.*, *98*, 8–15. doi: 10.1016/j.compbiomed.2018.05.011.

18. Heffernan, R., Yang, Y., Paliwal, K., & Zhou, Y., (2017). Capturing non-local interactions by long short-term memory bidirectional recurrent neural networks for improving prediction of protein secondary structure, backbone angles, contact numbers, and solvent accessibility. *Bioinformatics*, *33*(1)8, 2842–2849. doi: 10.1093/bioinformatics/btx218.

19. Mnih, A., & Salakhutdinov, R. R., (2008). Probabilistic matrix factorization. In: *Proc. Adv. Neural Inf. Process. Syst.* (pp. 1257–1264).

20. Koren, Y., Bell, R., & Volinsky, C., (2009). Matrix factorization techniques for recommender systems. *IEEE Comput.*, *42*(8), 30–37.

21. Zhong, J., & Li, X., (2010). Unified collaborative filtering model based on combination of latent features. *Expert Syst. Appl.*, *37*(8), 5666–5672.

22. Wang, C., & Blei, D. M., (2011). Collaborative topic modeling for recommending scientific articles. In: *Proc. 17th ACM SIGKDD Int. Conf. Knowl. Discovery Data Mining* (pp. 448–456).

23. Wang, H., Wang, N., & Yeung, D. Y., (2015). Collaborative deep learning for recommender systems. In: *Proc. 21st ACM SIGKDD Int. Conf. Knowl. Discovery Data Mining* (pp. 1235–1244).

24. Li, S., Kawale, J., & Fu, Y., (2015). Deep collaborative filtering via marginalized denoising auto-encoder. In: *Proc. 24th ACM Int. Conf. Inf. Knowl. Manage* (pp. 811–820).

25. Ying, H., Chen, L., Xiong, Y., & Wu, J., (2016). Collaborative deep ranking: A hybrid pair-wise recommendation algorithm with implicit feedback. In: *Proc. 20th Pacific Asia Conf. Knowl. Discovery Data Mining* (pp. 555–567).

26. Dong, X., Yu, L., Wu, Z., Sun, Y., Yuan, L., & Zhang, F., (2017). A hybrid collaborative filtering model with deep structure for recommender systems. In: *Proc. 31st AAAI Conf. Artif. Intell.* (pp. 1309–1315).

27. Sánchez-Escalona, A. A., & Góngora-Leyva, E., (2018). Artificial neural network modeling of hydrogen sulphide gas coolers ensuring extrapolation capability. *Math. Model. Eng. Problems*, 5(4), 348–356.

28. Lamia, S., Ben, S. O., Khalil, C., & Ben, H. A., (2015). Breast cancer ultrasound images' sequence exploration using BI-RADS features' Extraction: Towards an advanced clinical aided tool for precise lesion characterization. *IEEE Transactions on Nanobioscience, 14*(7).

29. Kingma, D. P., & Welling, M., (2013). *Auto-Encoding Variational Bayes*. [Online]. Available at: https://arxiv.org/abs/1312.6114 (accessed on 18 December 2020).

30. Li, X., & She, J., (2017). Collaborative variational autoencoder for recommender systems. In: *Proc. 23rd ACM SIGKDD Int. Conf. Knowl. Discovery Data Mining* (pp. 305–314).

31. Gubern-M'erida, A., Michiel, K., Ritse, M. M., Robert, M., & Nico, K., (2015). Breast segmentation and density estimation in breast MRI: A fully automatic framework. *IEEE Journal of Biomedical and Health Informatics, 19*(1).

32. Xingyu, L., Marko, R., Ksenija, K., & Konstantinos, N. P., (2019). *Discriminative Pattern Mining for Breast Cancer Histopathology Image Classification via Fully Convolutional Autoencoder, 7*. doi: 10.1109/ACCESS.2019.2904245.

33. Arjun, P. A., Alan, J. G., Junmei, C., Krishna, R. K., Richard, M. W., Liewei, W., Zbigniew, T. K., & Ravishankar, K. I., (2018). Machine learning helps identify new drug mechanisms in triple-negative breast cancer. *IEEE Transactions on Nanobioscience, 17*(3).

34. Eleftherios, T., Georgios, C. M., Katerina, N., Konstantinos, D., Manos, C., Antonios, D., & Kostas, M., (2019). Extending 2-D convolutional neural networks to 3D for advancing deep learning cancer classification with application to MRI liver tumor differentiation. *IEEE Journal of Biomedical and Health Informatics, 23*(3).

35. Bo, T., Ao, L., Bin, L., & Minghui, W., (2019). *CapSurv: Capsule Network for Survival Analysis with Whole Slide Pathological Images, 7*. doi: 10.1109/ACCESS.2019.2901049.

36. Syed, U. A., Shamim, H. M., Ghulam, M., Musaed, A., & Md. Abdur, R., (2019). *Cognitive Smart Healthcare for Pathology Detection and Monitoring, 7*. doi: 10.1109/ ACCESS.2019.2891390.

37. Musaed, A., & Ghulam, M., (2018). *Voice Pathology Detection Using Deep Learning on Mobile Healthcare Framework, 6*. doi: 10.1109/ACCESS.2018.2856238.

38. Qiuling, S., Fenglong, M., Ye, Y., Mengdi, H., Weida, Z., Jing, G., & Aidong, Z., (2018). Deep patient similarity learning for personalized healthcare. *IEEE Transactions on Nano Bioscience*.

39. Xiaoyi, D., & Feifei, H., (2019). Collaborative variational deep learning for healthcare recommendation. *IEEE Access*.

40. Musaed, A., & Ghulam, M., (2019). Automatic voice pathology monitoring using parallel deep learning for smart healthcare. *IEEE Access*.

41. Amyar, A., Ruan, S., Gardin, I., Chatelain, C., Decazes, P., & Modzelewski, R., (2019). 3D RPET-NET: Development of a 3D PET imaging convolutional neural network for radiomics analysis and outcome prediction. *IEEE Transactions on Radiation and Plasma Medical Sciences, 3*(2).

42. Chunxue, W., Chong, L., Naixue, X., Wei, Z., & Tai-Hoon, K., (2018). A greedy deep learning method for medical disease analysis. *IEEE Access, 6*.

43. Zhennan, Y., Yiqiang, Z., Zhigang, P., Shu, L., Yoshihisa, S., Shaoting, Z., Dimitris, N. M., & Xiang, S. Z., (2016). Multi-instance deep learning: Discover discriminative local anatomies for body part recognition. *IEEE Transactions on Medical Imaging, 35*(5).

44. Marwan, A. A., (2019). Skin Lesion classification using convolutional neural network with novel regularizer. *IEEE Access, 7*.

45. David, T., Maschenka, B., Otte-Höller, I., Rob, V. D. L., Rob, V., Peter, B., Carla, W., et al., (2018). Whole-slide mitosis detection in H&E breast histology using PHH3 as a reference to train distilled stain-invariant convolutional networks. *IEEE Transactions on Medical Imaging, 37*(9).

46. Andreas, H., (2016). *Machine Learning for Health Informatics* (pp. 1–24). LNAI 9605. doi: 10.1007/978-3-319-50478-0 1, Springer.

47. Dimitrios Kollias, Athanasios Tagaris, Andreas Stafylopatis, Stefanos Kollias & Georgios Tagaris, (2017). Deep neural architectures for prediction in healthcare. *Complex Intell. Syst.* doi: 10.1007/s40747-017-0064-6, Springer.

CHAPTER 2

AN OVERVIEW OF CONVOLUTIONAL NEURAL NETWORK ARCHITECTURE AND ITS VARIANTS IN MEDICAL DIAGNOSTICS OF CANCER AND COVID-19

M. PAVITHRA,[1] R. RAJMOHAN,[2] T. ANANTH KUMAR,[1] and S. G. SANDHYA[2]

[1]Assistant Professor, Department of CSE, IFET College of Engineering, Gangarampalaiyam, Tamil Nadu, India,
E-mails: pavimuthu27@gmail.com (M. Pavithra),
ananth.eec@gmail.com (T. A. Kumar)

[2]Associate Professor, Department of CSE, IFET College of Engineering, Gangarampalaiyam, Tamil Nadu, India,
E-mails: rjmohan89@gmail.com (R. Rajmohan),
sgsandhyadhas@gmail.com (S. G. Sandhya)

ABSTRACT

The term artificial intelligence (AI) itself represents that machine with minds, i.e., the machine will able to think like a human and behave as a human-based on the situation. Later a field of AI called deep learning (DL) was developed. DL exactly tries to mimic the human brain, i.e., it is modeled based on the functions of the neocortex. DL was modeled to understand the functions of the neurons in the human brain and how the neurons stimulate the brain to think [1]. In addition, the results of DL are massive in real-time like speech recognition, handwriting recognition, disease prediction, etc. DL contains two models one is generative architectures, and the other is discriminative architectures. Generative DL model adapts unsupervised learning, the algorithm itself extract the relationships among the data. The discriminative DL model adapts the techniques of classification and regression. This model is based on the supervised learning algorithm, i.e., with the help of some training data, the algorithm will learn, and based on the training, it classifies the further testing data [2]. The best examples of discriminative DL models include deep neural networks (DNN), convolutional neural network (CNN), and recurrent neural network (RNN). This chapter explores the basic architectures of CNN along with its variants in cancer diagnosis and COVID-19.

2.1   INTRODUCTION

CNN is initially proposed for 2-dimensional data processing like images and videos and also texts. In recent years CNN to be proved as an efficient learning algorithm for image recognition. ConvNet (CNN) is one kind of image processing algorithm that is used to extract features from images [2]. It extracts features based on the edges (horizontal and vertical) and color. CNN is designed with three layers they are convolutional layers, pooling layers, and fully connected layers (FC layer).

2.2   SIMPLE STRUCTURE OF CONVOLUTIONAL NEURAL NETWORK (CNN)

2.2.1   ONE DIMENSIONAL STRUCTURE OF CNN

Let us consider user want to develop a language detection neural network model to detect the language spoken by a human. The input is an audio sample extracted at different points of time. Assume the audio samples collected at different points of time as A1, A2, A3,…,An.

Here the audio samples are considered as the set of neurons, and each input is connected to each other input through a FC layer is illustrated in Figure 2.1. Based on the similar properties among the neurons, i.e., the audio frequency may be the same for two or more neurons, create a set of neurons, N. N groups each neuron based on similar features called convolutional layer [2]. One advantage of the convolutional layer is, the user can add one or more convolutional layers to extract the features from the neurons. This layer fed the input to a FC layer.

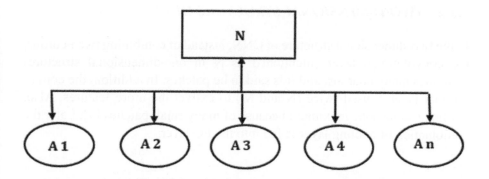

FIGURE 2.1 One-dimensional structure of CNN.

After that pooling layer is introduced, which is closely related to the convolutional layer, it represents the data at a higher level and also reduces the parameter of the convolutional layer by combining it. Generally, two pooling functions are specified: average pooling; and maximum pooling [3]. Average pooling specified the average value from each block and maximum pooling specifies the maximum value from each block. The pooling layer is illustrated in Figure 2.2.

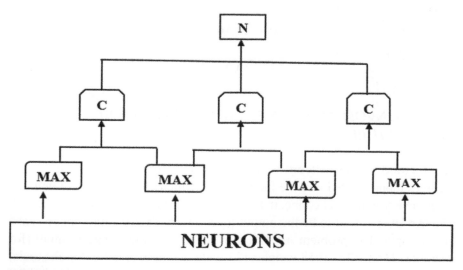

FIGURE 2.2 Pooling layer.

2.2.2   *TWO-DIMENSIONAL STRUCTURE OF CNN*

In the two-dimensional structure of CNN, instead of combining two neurons, the convolutional layer (mentioned as N in one-dimensional structure) combines many neurons, and it is said to be patches. In addition, the convolutional layer is used twice (N and M) to extract the more features. Also, pooling can be done maximum because of many small patches [4]. Later the combination of pooling layer is fed into the FC layer.

2.2.3   *CONVOLUTIONAL NEURAL NETWORK (CNN) FORMALIZING*

Formalizing the neurons based on the architecture used is considered as important in the neural network. To formalize the CNN, consider the one-dimensional structure of CNN. The one-dimensional equation for CNN is represented in Eqn. (1) here A1. An -is considered as input and Z1. Zn is considered as output. Now the output can be described in terms of input as follows:

$$Zn = N\,(An,\,An + 1) \tag{1}$$

i.e.:

$$Z0 = N\,(A0,\,A1);$$

$$Z1 = N\,(A1,\,A2).$$

2.2.4   *CONVOLUTIONAL NEURAL NETWORK (CNN) ARCHITECTURE*

Feedforward neural network (FNN) is made up of neurons. Each neuron receives input and fed into the hidden layer, which contains a set of neurons where every neuron is connected to every other neuron in the network [5]. The last layer is said to be the output layer which classifies the image based on the input. The problem in FNN is if the size of the image is small (for example, $23{\times}23{\times}2 = 1058$ weights (23 wide, 23 high, and 3 color channels) it is easy for the fully connected neurons to classify images in the first hidden layer. However, the size of the image is large (for example, $300{\times}300{\times}4 =$

360,000 weights) it is difficult for FNN to manage the large weight using a single hidden layer [6]. To overcome this, CNN architecture is introduced. CNN algorithm compresses the image using three layers and produces the output.

2.3 LAYERS IN CONVO NET

As mentioned already, CNN is designed by using three layers called convolutional layer, FC layer, and pooling layer. The convolutional layer combines the input neurons into a set of patches and the pooling layer performs pooling function to extract the higher level of abstraction from the patches and fed it into the FC layer [5]. Let us discuss the working of the individual layer separately.

2.3.1 CONVOLUTIONAL LAYER

As the name indicates, it is one of the important building blocks in Convo Net. The human brain is a very complicated and powerful machine, which captures multiple sets of images per second and processing the image without realizing how it was done. However, the same case does not suit for the machine. The machine must know how to process the image to extract some output [7]. Generally, every machine considers an image as an arrangement of the pixel, if the arrangement of the pixel is changed, the image also changed. Therefore, machines recognize the image in terms of pixels. The convolutional layer takes the image as an arrangement of pixels is input and performs some filter operations using hyperparameters to produce an output. Consider an image of size 6×6 as an input. Now define the weight matrix for an input image as 3×3 to extract the features from the input image. The architecture for CNN is illustrated in Figure 2.3.

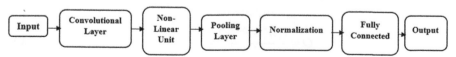

FIGURE 2.3   Architecture of convolutional neural network.

The 3×3 weight matrix should run across the input matrix horizontally to extract the features from the input image. In addition, convert the size of the input image into a 4×4 matrix. The size of the input image is reduced by using a weight matrix filter, it acts as a filter to extract the color code, reduce the unwanted noise from the input image, etc. It is recommended to use two or more convolutional layers depend on the size of the image because by going deeper, it extracts the multiple features which lead to the correct prediction [8]. However, going deeper will leads to complex the situation to extract the filters.

2.3.2 TUNING HYPER PARAMETER IN CONVOLUTIONAL LAYER

In the convolutional layer, two hyperparameters are used to control the spatial arrangement of the image. They are stride, padding, and dimensions. Stride controls the filter convolves with the input pixels; this is used to reduce the image quality using the filter [7]. The filter size is not standardized one based on the volume of the dataset, the size of the filter might vary. If the size of the stride is 1, then the movement of the pixel is one at a time. If the size of the pixel is 2, then the movement of the pixel is w at a time. Depends on the volumes of the dataset, the size of the strides may be varied [8]. The output of strides is to produce a smaller volume of data compared to input volume. If the stride value is increased continuously, then the size of the input image gets smaller, to solve this problem zero padding method is introduced [8]. Here zero can be padded across the border of the image, which is used to maintain the spatial size of the output image as same as the input image it is said to be the same padding [8]. If the size of the stride is 1, then use the below formula to calculate the zero-padding:

$$Zero\ padding\ = \frac{((F-1))}{2} \qquad (2)$$

where; F = size of filter.
    The size of the output is calculated using the below formula:

$$O = \frac{((W-K+2P)}{2}+1 \qquad (3)$$

where; O is the output height/length, W is the input height/length, K is the filter size, P is the padding, and S is the stride.

2.3.3 NONLINEAR LAYER

To apply the nonlinearity functions in CNN, the activation function is introduced immediately afterward the Convolutional layer. The convolutional layer performs only the linear operations like element-wise multiplication and summation, so next to the convolutional layer, some of the activation functions are used to improve the nonlinearity [8, 9]. Mostly ReLU (rectified linear unit) is used as a nonlinearity function in CNN. The formula to represent the Relu is represented in the below equation:

$$F(x) = max\ (0,x) \tag{4}$$

The main purpose of the Relu layer is to convert the all-negative value in the convolutional layer is to 0, this helps to overcome the problem of vanishing and gradient problem which is faced by most of the neural network's algorithms [8]. The Relu layer changes the network into nonlinear without affecting the accuracy of the network.

2.3.4 POOLING LAYER

As the name indicates, this layer tries to pool the values by performing certain functions, and also the pooling layer is said to be downsampling [9]. The pooling layer is also used to reduce the spatial dimensions from the output of the nonlinear layer and also minimize the computation of the network; hence it prevents the occurrence of overfitting. Generally, three types of pooling functions are there max pooling, average pooling, and L2 norm pooling [8, 9]. Max pooling is the most common method used in CNN, and the common size of filter for max-pooling is 2×2 is applied for the output of the Relu layer, the max-pooling operation is illustrated in Figure 2.4.

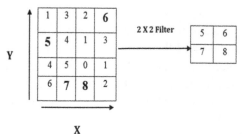

FIGURE 2.4   Max pooling.

2.3.5   *FULLY CONNECTED LAYER (FC LAYER)*

Normalization functions like drop out are used by the network to avoid the overfitting. This can be used only in the training phase to tune the network not in the testing phase. The idea behind normalization is to set all the activations as 0 because it causes the overfitting problem in the network [9]. The output of the normalization layer is fed into the FC layer, where the neurons are fully connected with all the activation functions in the previous layer. FC layer flatten the 3-dimensional data into the vector and perform element-wise operations to classify the images. The formula for a FC layer is represented as follows:

$$FC = A \ (Wi + b) \tag{5}$$

where; A = Is the activation function which is mostly SoftMax, Relu; W = weight vector; i = input vector; and b = bias.

2.4   CASE STUDY: APPLICATIONS OF CNN IN MEDICAL FIELD

Nowadays, deep learning (DL) algorithm recreating the medical field because of its ability to predict the risk of diseases in the future, it plays a vital role in many diseases like diabetes, heart disease, etc. Cancer is one of the major diseases in today's world and it's still hard to find the reason for cancer and even in most of the cases, it is very critical to identify the disease [10, 11]. Particularly breast cancer is common among women's and there is a risk of death for women if they are not undergoing any diagnosis. The screening technique available for breast cancer is mammography it results in an x-ray image of the breast with a low dose of radiation, it is only an available technique to identify breast cancer before patients felt the tumor disease [10]. Generally, two types of mammography tests have been followed by medical professional they are screening mammography and diagnostic mammography. The screening method is preferred for the patients who don't have any signs of illness and the diagnosis method is recommended for the patients who have some primary symptoms like breast thickening, nipple discharge, and change in size and shape of the breast. The two techniques of mammography that can be used to capture the breast images of the patient are film screen mammography; it is a combination of black and white images with a film roll, and using this film, the medical professional has to screen the patients. Another method is full-field digital mammography (FFDM), which is fully based on a computer, and the image of the breast can be captured and record directly into the computer. The professional will able to refer the images later, and there is the possibility to transfer the mammographic results electronically [11]. FFDM is an efficient way to identify the risk of cancer, but it is not available in worldwide and also very expensive compared to film-screen mammography. Computer-aided design (CAD) is one method of examining the mammographic images where it uses the computer software to distinguish the mass tissue from normal tissue. Even though mammography is the only way to detect breast cancer, it has some limitations. It cannot determine the malignant and non-malignant cancer disease. CNN and its variants are used to detect and classify and diagnosis of breast cancer [11]. Many researchers have been reported related to CNN in the field of breast cancer based on the mitotic rate and classify the cancerous and non-cancerous cells.

2.4.1 CONVOLUTIONAL NEURAL NETWORKS (CNN) TO CLASSIFY THE BENIGN AND MALIGNANT CELL

Cancer is the abnormal growth of the cells in the human body or a small lump, also said to be a tumor. The abnormal growth of the cells occurs in any part of the human body like lungs, skin, breast, etc. Generally, medical professional diagnoses and classifies the cancer cells into two types they are benign and malignant cells [11]. The benign cell is also said to be a non-cancerous cell because it cannot occupy or spread to its surrounding tissue from the affected area, so the spreading risk of cancer from one tissue to another tissue is less. However, in rare cases, the benign tumor can spread slowly to the surrounding tissue if the medical professionals identify it at an early stage and remove it in the initial stage itself means they do not grow. Another type of tumor is premalignant tumor these cells are neither malignant nor benign, but it can change into the cancerous cell. The last type of tumor is malignant cell invades or spread to its surrounding tissue and increases the risk of cancer. So initially, it is important to classify the cell for further treatment [10, 11]. Manual analysis to classify the cell using mammographic images has existed, but in most of the cases, it is difficult for radiologists to differentiate the cells. CNN algorithm classifies the cells computationally with better accuracy. Researchers proposed different variants of CNN algorithms to differentiate the cells like breast cancer detection using CNN (BCDCNN) and mammogram classification using CNN (MCCNN).

Consider a mini-MIAS (Mammographic Image Analysis Society) database (DB), which is a collection of mammographic images of breast cancer [11]. Mini-MIAS contain the low size images compared to the original image, i.e., if the original size of an image is 200-micron pixel edge means the reduced size of an image is 50-micron pixels. Mini-MIAS DB is labeled DB, every image in the DB are labeled with the types of cell tissue, characteristics of the tissue. The sample of mammographic tissue is illustrated in Figure 2.5.

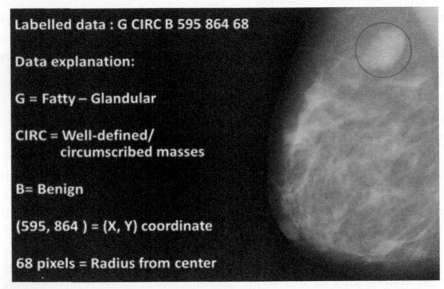

FIGURE 2.5   Labeled mammographic image.

The above image describes the sample labeled mammographic image where benign defines the type of cell tissue, G defines the characteristics of the cell, x, and y coordinates represent the abnormality center. Mostly mammographic images are large even though the original size is compressed into 50 microns [11]. Therefore, it is hard for the CNN algorithm to predict the accurate result also the algorithm spends a lot of time in computation time and to remove noise. To increase the accuracy and reduce the computation time, the image has to be crop down the important tissue from the mammographic DB. The below image (Figure 2.6) illustrates the cropped image of CNN classifier.

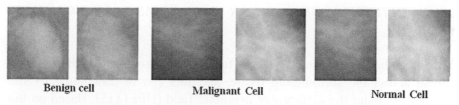

Benign cell   Malignant Cell   Normal Cell

FIGURE 2.6   Different kinds of cancerous cell.

The above diagram shows the cropped images of the different cells, which include normal patient's tissue also to differentiate it from the cancerous cell. As mentioned earlier the CNN architecture contains three layers the 48×48×1 mammographic image will give as the input to the network, the hidden layer contains two or more layers like convolutional layer, pooling layer, and FC layer [10]. Initially, the input was fed into the convolutional layer (5×5) with filter size 3×3. The next layer is nonlinear; here, Relu is used as an activation function to introduce the nonlinearity in the network. Late this nonlinear output is fed into the pooling layer; CNN uses a max-pooling mechanism to perform pooling operation. The size of the max-pooling filter is 2×2; it reduces the processing time and size of the image. CNN uses a backpropagation algorithm to update its weight and bias also to avoid the vanishing and exploding gradients problem. During the training phase, the loss rate and error rate are calculated using a backpropagation algorithm, by using this the model tries to improve its accuracy by reducing the calculated error and loss rate [13, 14]. The overall architecture for CNN in breast cancer detection is illustrated in Figure 2.7. The third layer is fully connected, each neuron in this layer is connected nodes in the previous layer, and this FC layer is also called a decision-making layer. Here, the layer has to differentiate 3-cells 0-normal cell, 1-benign cell, and 2-Malignant cell.

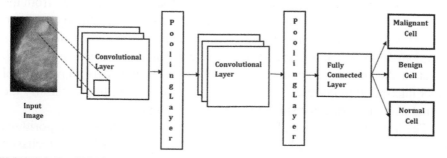

FIGURE 2.7   CNN based breast cancer detection model.

2.4.2   DETECTION OF MITOSIS RATE IN BREAST CANCER USING CNN

Mitosis is the cell division, i.e., how fast the cancerous cell is divided and spread across the surrounding tissue, and this can be calculated by the pathologist under the high-power magnetic field (HPF) [15]. Based on the mitotic rate, the aggressive of the disease can be calculated by the medical professional. Mitotic rate counting is a laborious work and consumes more time to identify the rate because it is a fully manual process; pathologist needs to calculate the rate under the microscopic analysis. Pathologist scales the mitotic rate count from 1 to 3, where '1' represents the mitotic count is less than 5 points under 10 HPF, '2' represents 5 to 10 points under 10 HPF, and at last '3' represents greater than 10 points under 10 HPF. The risk of breast cancer for 1 mitotic point is low (differentiate the cell easier) and 3 mitotic point is high (poor differentiation between cells) [16]. However, the mitotic rate calculation needs manual power and a time-consuming process to overcome it a computerized method has been proposed by researchers to calculate the mitotic rate using CNN algorithms.

2.4.2.1   METHODOLOGY FOR THE ANALYSIS OF MITOTIC RATE USING CNN ALGORITHM

To identify the mitotic rate, collect the biopsy slide images of breast cancer. The dataset available in open source like BreCaHAD (breast cancer histopathological annotation and diagnosis dataset) and MITOSIS ATYPIA 14, etc., are labeled images [17]. Generally, the initial step is the removal of noise and segmentation of the images. The data collected from open source DBs are higher in size so, it is important to crop the images into fixed-size before feeding it into the CNN algorithm. In addition, the images collected from the open-source DB follow RGB color space; this RGB space can be converted into YUV space and normalized the value of mean (0) and variance (1). As mentioned already, the CNN architecture contains three main layers [17]. Most of the mitosis detection CNN algorithms follow two convolutional layer max-pooling layer and FC layer. DL algorithms predict or classify the output using two stages testing phase and training phase.

The training phase is said to be the learning phase where the algorithms learned from the data used for training, in CNN algorithm SGD (stochastic gradient descent) have been used for training the dataset and to reduce the loss rate. The testing phase is used to either predict or classify the data based on the training; here, the algorithm needs to classify the mitotic cell and non-mitotic cell. The probability threshold fixed for the testing phase is 0.58 to identify the mitotic cell. If the probability threshold is greater than or equal to 0.58 means, then the cell is mitotic; it is calculated based on the nuclei. The architecture of Mitotic rate detection using CNN is illustrated in Figure 2.8. The biopsy image is considered as an input image, and after pre-processing (removal of noise), the candidate segmentation is done. The segmentation defines the conversion of RGB color space to a blue ratio image to highlight the nuclei [18, 19]. The next step is blob analysis, where the unwanted parts of the images are eliminated. Moreover, the structure is very small after removing the unnecessary parts. The CNN architecture takes this as input to classify the mitotic and non-mitotic cells. Convolutional Layer contains the 98-filter size with nonlinear activation function with max-pooling layer and the consecutive convolutional layer contains 384 filter size with one nonlinear activation function mostly Relu with max-pooling layer [20]. Dimensionality reduction is performed to increase the accuracy of the classification, and finally, the classification of cancerous and non-cancerous cells has been carried out. This is the general architecture procedure of mitosis cell detection for breast cancer, it may vary depends on the algorithm and datasets.

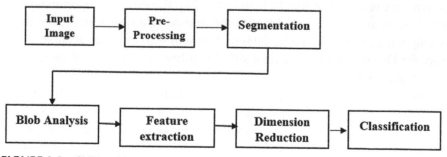

FIGURE 2.8   CNN architecture to classify mitotic cells.

2.4.3 INTRODUCTION OF COVID-19

Coronavirus, also known as COVID-19, is a pandemic disease in the current situation, which begins at the end of 2019 in China. The outbreak of the disease is worldwide, and the affected cases are around 3 million reported by the World Health Organization (WHO). Proper vaccination for COVID-19 has been still researched by many researchers worldwide. The only diagnosis method available for COVID-19 is RTPCR (reverse transcription-polymerase chain reaction), Kit. However, this kit does not support for the early detection of diseases and treatment [21]. In addition, COVID-19 is said to an asymptotic disease, the nature of coronavirus is it shows the symptoms to the person after 13–17 days it enters into the victim. However, 80% of cases are asymptotic cases reported worldwide, it does not cause any symptoms, but the test result is positive. Therefore, it is difficult to identify COVID-19 disease. Recently, many researchers have been reported by researchers worldwide to identify the COVID-19 at the early stage using the advancement of artificial intelligence (AI). The only thing required by AI algorithms is data. Based on the data trained to the neural network, it gives the test result more accurately. CNN one of the DL algorithms which train the networks based on the images [21]. Within the short period, many works are reported based on CNN and COVID-19. CNN can be used to detect the coronavirus from the chest x-ray (CXR) images of the patient, and also it will able to differentiate the coronavirus from pneumonia. The impact of COVID-19 is worldwide and day by day the death rate and infected rate of the disease has been increased rapidly. Still, there is no vaccine, and methods to detect the disease at an early stage have not been proven even though many kinds of research have been reported. The disease cannot be identified at early stages because of the lack of symptoms, and even many asymptotic patients are reported around 80% in India [21]. Researchers report many DL and AI-driven algorithms to identify the disease at an early stage, but there is only a minimum amount of datasets that will lead to better accuracy. It is easy for DL algorithms to train a small number of datasets and predict the testing accuracy.

Medical professionals use hydroxychloroquine and azithromycin to treat COVID-19 infected patients worldwide. Still, the spreading of disease and death rates increased. In addition, lockdown is the only solution to avoid the spreading of disease, even patients are re-admitted in many countries once they are cured, and again, they got affected within 50 days.

COVID-19, it is a pandemic disease announced by the WHO in 2020. Within four months, the virus spread worldwide and affects the survival of humans and animals. Researchers doing many kinds of investigations to control the spreading and identifying the vaccine for COVID-19. On the technological side, researchers try to predict the diseases in the early stages using some techniques or algorithms. Some of the reported works of coronavirus are as follows:

1. **AI-driven tools for coronavirus outbreak:** *Need for active learning and cross-population train/test models on multitudinal/multimodal data [22]:* This chapter is about active learning in AI. In the current situation, active learning is very important because there is no standard dataset for COVID-19 till now. The range and features of data may vary day today. Therefore, based on the minimum data, they try to predict the disease using active learning and DL [22]. Based on the initial dataset, they try to make decisions with the help of AI-driven tools in parallel with experts in the field. However, the result is accurate, at least with some more data.

2. **CT radionics can help screen the coronavirus disease 2019 (COVID-19):** A preliminary study Mengjie FANG et al.: In this chapter, the radionics method has been used. It uses the radiometric methods to screen the COVID-19 patients from CT images. With the help of 79 patient's data (50 pneumonia and 29 COVID-19) and 77 radiometric features have already been extracted from the segmentation of lungs using CT images. By using a support vector machine (SVM) and statistical technique, they try to classify other pneumonia from COVID-19 [23]. Again, the dataset is minimum and collected from a single hospital, and it is not enough to train the algorithm and predict the results.

3. **COVID-19 outbreak prediction with machine learning [23]:** This study reports the comparative studies of machine learning algorithms and soft computing. Due to the lack of data, standard models shows low accuracy for long-term prediction. The results of multi-layer perceptron (MLP) and ANFIS reported high generalization ability for long-term prediction. They do not report anything about the mortality rate; it is important to estimate the mortality rate to estimate the number of patients and estimate the number of required beds.

4. **Coronavirus optimization algorithm:** A bioinspired metaheuristic based on the COVID-19 propagation model F. Martínez-Álvarez, G. Asencio-Cortés et al. [24]: This chapter is about the spreading of coronavirus across the word. The detailed study about how the disease is spreading the community and how to stop the spreading of diseases using corona optimization algorithms along with DL models.

5. **Geographical tracking and mapping of coronavirus disease COVID-19/severe acute respiratory syndrome coronavirus 2 (SARS-CoV-2) epidemic and associated events around the**

world: How 21st century GIS technologies are supporting the global fight against outbreaks and epidemics [25]. This research is about the tracking of the COVID-19 patients using GIS technologies through mobile applications and alert the peoples through the applications. The application calculates the distance of COVID-19 patients from the user's location to alert the users. Many types of research have been based on the early prediction of the diseases using DL algorithms and AI-driven tools.

2.4.4   DETECTION OF COVID-19 BASED ON THE CHEST X-RAY (CXR) IMAGE USING CNN

Let's discuss how the COVID-19 can be detected using the CNN algorithm. The dataset considers for this work is from GitHub. It contains the CXR images of the patients having COVID-19 syndrome, ARDS syndrome, and SARS. CNN algorithm takes the input of COVID-19 images along with normal patient's images collected from Kaggle [26]. In addition, the images have to be cropped and resized into the fixed-size suits for the CNN algorithm (224×224). Most of the study uses variants of CNN algorithms like ResNet, AlexNet, and Inception V3 to differentiate COVID-19 CXR images from normal CXR images.

The above architecture (Figure 2.9) illustrates the CNN model for COVID-19. The input image is fed into the per trained CNN networks. Inception V3 is one classifier of CNN where it uses multiple filters at a single level. The inception model is proposed to choose the correct filter size based on the locomotion of the image. It uses different filters like 3×3 filter, 2×2 filters, etc., at a single level along with the max-pooling layer. Different versions of inception architecture are developed up to V3 [26]. Each version is differing from one another. Residual network (ResNet) is one variant of a CNN; it is mainly introduced to overcome the problem of vanishing gradient and exploding gradient. The main idea behind every neural network algorithm is to going deeper into the model (by increasing the number of hidden layers) will increase the accuracy of the model. However, this leads to vanishing and exploding gradient problem. By going deeper, the model forgets about the previous layers, this can be avoided with the help of the ResNet. In COVID-19 classification, the Inception and ResNet algorithms are pre-trained with the help of the sample COVID CXR images and normal X-ray images. The new set of input images is fed into this pre-training model the output of this model is fed into the pooling layer. The pooling layer used here is the global average-pooling layer; it performs the average pooling from the previous layer. It reduces the image size to improve the accuracy based on the average pooling. Average pooling chooses average values from the previous layer; if the filter size is 2×2, then the filter is applied to every layer in the pre-trained model and extracts the average value from the filter. In addition, the last layer is a FC layer where every node in the network is connected to every other node in the previous layer. The FC layer uses two types of classifiers they are Relu and Softmax for COVID detection [26]. In addition, the final output is the classifier classifies the COVID-19 CXR image and Normal CXR images independently. If new input is fed into the model means it will able to differentiate the COVID image and it is the fast and efficient process to identify the COVID-19 patients. CNN and its variants re-evaluate the medical sector, particularly in the field of medical imaging like liver segmentation, cervical cancer, etc.

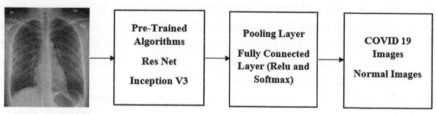

FIGURE 2.9   Architecture for COVID-19 x-ray image classification.

2.5   CONVOLUTIONAL NEURAL NETWORKS (CNNS) VARIANTS BASED ON THE REGION

2.5.1   R-CNN

R-CNN stands for region CNN, one of the variants of CNN introduced in 2015. This model combines the region with CNN to detect and classify the object. Nowadays, R-CNN is used in various applications like medical imaging, ship detection, etc., for detecting objects. R-CNN takes an input image, and it extracts a bottom-up region proposal from an input image; for example, assume the bottom-up region for the input image is 3000. Now R-CNN algorithm tries to compute features for every single bottom-up region extracted in the previous step with the help of the large CNN network, and the last step is to classify the regions using the machine learning classifiers like SVM, etc. R-CNN follows three modules, the initial module is region proposal and followed by a convolutional network and the third module is a class-specific linear classifier [27]. The region proposal model detects the individual region of the object accurately. The object recognition algorithm recognizes the object and label the object in the input. For example, if the laptop image is fed as an input to the object recognition algorithm means it takes the whole image as input and returns the labeled output. However, the object detection algorithm takes the whole image as input and divides them into regionally, and returns the individual region of the output with its coordinates. For example, consider the previous laptop image as an input, object detection algorithm detects every individual region in a laptop like a keyboard, mouse, screen, etc., along with its coordinates, i.e., height and width of the individual object [27]. The region proposal model mostly used a selective search algorithm to detect the object using its region. Selective search algorithm performs clustering operation, and the term clustering defines the grouping of similar character like this the algorithm combines the pixels of the images if it has similar characters like color, texture shape and size of the image. The region proposal method does not meet the goal for a perfect segmentation, and the main goal is to overlap many regions for prediction using over segmenting method. Later the over segmenting image is considered as an input by the selective search algorithm, and then it proceeds with the below steps 1. Create and add bounding boxes 2. Cluster the adjacent grouping boxes based on the similarity of adjacent boundary boxes. As mentioned earlier, the similarity index between adjacent regions can be calculated based on the texture of the image, the color of the image, size, and shape of the image. In addition, the selective search is performed by using the hierarchical clustering algorithm (HCA). Initially, the input image performs segmentation by using the clustering sub-segmented images are grouped; this is the simple working procedure of the HCA algorithm. Again, here grouping is done by measuring the similarity index between the adjacent region. The texture similarity can be calculated using the below equation:

$$T_{(texture)} (X_a, X_b) = \sum_{p=1}^{m} \min (U_a^{p} U_b^{p}) \qquad (6)$$

where, X_a and X_b are two regions; U_a and U_b are histogram regions; and p is a descriptor of text.

Like this shape and size similarity has an individual formula to calculate the similarity index between the adjacent regions. The next step in RCNN is feature extraction. CNN algorithm performs the feature extraction before fed into the CNN algorithm, and the region must be wrapped into the fixed size. The fixed size for the CNN algorithm in region CNN is 227×227 pixels [26]. CNN performs classification and feature extraction in Region CNN using AlexNet (One of the variants of CNN). It contains five consecutive convolutional layers with two FC layers. AlexNet was initially proposed in the Caffe model. The wrapped RGB input is fed into the AlexNet to extract the feature, and finally, SVM linear classifier is used to classify the input image. The overall architecture for RCNN is illustrated in Figure 2.10.

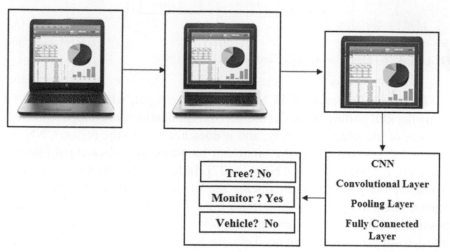

FIGURE 2.10   Detection of an object through region CNN.

Region CNN addresses some limitations like only it takes fixed regions to classify the image and it is very slow and time-consuming even to test a single image and there is no learning occurred here because a selective search algorithm is the fixed one.

2.5.2   FAST RCNN

Fast RCNN tries to overcome the limitations of region CNN. Rather than extracting CNN vectors autonomously for every region proposal, Fast region CNN combined those into one CNN forward pass over the whole picture, and the region proposal shares this element lattice. At that point, a similar component network is fanned out to be utilized for learning the item classifier and the bounding box regressor. Taking everything into account, calculation-sharing velocities up R-CNN. The overall architecture of RCNN is illustrated in Figure 2.11. Fast region CNN performs an ROI pooling operation. ROI pooling stands for region of interest pooling, is particularly used to detect the object in the CNN.

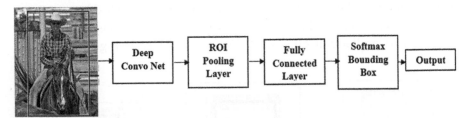

FIGURE 2.11   Detection of object through fast region CNN.

The major limitation of RCNN is fixed size input, which can be solved using the ROI polling layer. Depends on the applications and image collected, the size of the image may vary, and it does not fit into the region CNN, so before fed into the CNN, the input can be wrapped and changed into the fixed size (2000) in region CNN [26, 27]. In fast RCNN ROI pooling layer overcome the problem by producing the fixed-size feature. This fixed-size feature helps to map the non-uniform input image to the output image. It is possible because ROI performs max pooling on the input image to convert it into a fixed size. The difference between RCNN and Fast RCNN is it takes the whole image as an input to the CNN layer instead of segmenting. After extracting the feature from the input image, it maps the region based on the feature but RCNN takes regional segmented image as an input. After extracting the feature from an input image fast RCNN maps it to the size of the output layer using the ROI pooling layer. Finally, a SoftMax classifier is used to classify the image. Fast RCNN is computationally faster in both testing and training, but it is expensive and still uses a selective search algorithm.

2.5.3   FASTER RCNN

It is the combination of region proposal network (RPN) and fast RCNN developed to overcome the limitations of fast RCNN [26]. The working of faster RCNN is described as follows:

- Pre-train a CNN arrangement on picture grouping assignments.
- Adjust the RPN (locale proposition arranges) start to finish for the district proposition task, which is instated by the pre-train picture classifier. Positive examples have IoU (convergence over-association) > 0.7, while negative examples have IoU < 0.3.
- Slide a little n×n spatial window over the convolution include guide of the whole picture.
- At the focal point of each sliding window, we foresee numerous areas of different scales and proportions at the same time. A grapple is a mix of (sliding window community, scale, proportion). For instance, 3 scales + 3 proportions => k=9 stays at each sliding position.
- Train a fast R-CNN object identification model utilizing the recommendations produced by the current RPN.
- At that point, utilize the fast R-CNN system to introduce RPN preparation. While keeping the common convolutional layers, just adjust the RPN-explicit layers. At this stage, RPN and the recognition organization have shared convolutional layers.
- At long last, tweak the exceptional layers of Fast R-CNN.
- Stage 4–5 can be rehashed to prepare RPN and fast R-CNN, on the other hand, if necessary.

2.6   CASE STUDY: MEDICAL APPLICATIONS FOR REGION-BASED CNN AND ITS VARIANTS

2.6.1   DETECTION OF LIVER USING FASTER REGION CNN

Liver is one of the main functioning organs in the human body, which contains multiple blood vessels and the complex structured organ of the human body. Now a day's peoples are mostly affected by liver-related problems like liver cancer, etc. Therefore, for detection and diagnosis of liver-related diseases required proper segmentation of the liver. Based on the segmentation results, the medical professional will able to diagnose and treat the liver related disease [27]. Even surgical decisions are made by the doctor based on the segmentation process from the images of the liver obtained from CT scan X-ray, etc. Because of the complex structure and shape of the liver, it is difficult for medical professionals to segment the liver to make a decision, it is a time-consuming process. So digitalized liver segmentation method is proposed by many researchers based on the deep lab and CNN algorithms. Automated CT liver segmentation is in practice, but it faces some difficulties like the intensity of the liver is similar to its corresponding organs known as heart and intestine, and the size of the liver is not the same for everyone it may vary from one person to another person [28]. Therefore, the existing method cannot able to segment the liver efficiently. Faster region CNN detects the liver accurately by differentiating it with the neighboring organs having the same intensity. Initially, the images are taken from many open source DBs like LiverTox, etc., and then fed the whole image into the VGG to extract the features from the images. Then faster region CNN uses RPN to extract the region candidate from the whole image approximately 300 candidates have been extracted by the RPN algorithm. VGG model is used for large-scale images, now map the region proposal feature map to the last layer of the convolutional layer in VGG modeling. The extracted future map is different in size, so it needs to be resized; this is possible by using the ROI pooling layer. ROI pooling layer performs a max-pooling operation to map the size of the input layer to the size of the output layer [29, 31]. Finally, mapped input is fed into the FC layer and it results in the detection of the liver. The architecture for liver detection using faster RCNN is illustrated in Figure 2.12.

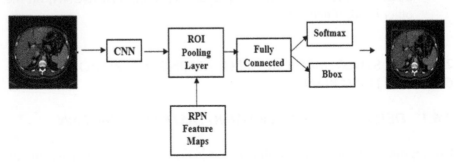

FIGURE 2.12   Detection of liver using faster RCNN.

The input image is a CT scanner image that contains the liver and its surrounding organs like heart, stomach, etc. To analyze the disease aggressively in the liver first, it is required to detect and differentiate the liver from other organs [30, 31]. The input image is fed into the faster region CNN, first, it extracts the features from the image and performs region proposal to split the image into a region and then ROI pooling is performed to map the input size with the output size, after mapping the resultant image is fed into the FC layer, by using SoftMax classifier it classifies the detected object from the input image using boundary box analysis method. Compared to other detection algorithms, it gives better accuracy, and it is computationally faster. Neural network algorithms are implemented using the Keras and Tensorflow platform. Anaconda is an open-source tool available to implement all kinds of neural network algorithms. Both CPU and GPU versions of algorithms can run in the anaconda platform with the help of Keras and TensorFlow. Tensorflow inbuilt many algorithmic packages of neural networks like LSTM CNN GRU, etc. In addition, Google Colaboratory (Colab) is available as an open platform with GPU capacity to execute the neural network algorithms.

2.7 CONCLUSION

AI is an emerging field which tries to recreate the machine through human intelligence. Researchers introduced many variants of CNN algorithms based on the application domain like cascaded CNN, CB CNN, etc. The convolutional network is the basic architecture, based on these different variants like YOLO (You Look Only Once), Faster RCNN, RCNN, AlexNet, GoogLeNet are developed. Not only in the field of medical imaging, CNN algorithms are used in multiple applications in real-time like ship detection, satellite communication, etc. This chapter explores the basic architecture of CNN along with Faster RCNN, RCNN, Fast RCNN based on RPN for medical imagining. CNN variants have an ability to classify the images and detect the risk of diseases.

KEYWORDS

- artificial intelligence
- computer-aided design
- convolutional neural networks (CNN)
- deep learning
- deep neural networks
- discriminative architecture

REFERENCES

1. Bengio, Y., (2009). Learning deep architectures for AI. *Foundations and Trends in Machine Learning, 2*(1), 1–127. Also published as a book, Now Publishers.

2. Abdelrahman, H., & Anthony, P., (2016). *A Study on Deep Learning*. Slide Share.

3. Arel, I., Rose, D. C., & Karnowski, T. P., (2010). Deep machine learning: A new frontier in artificial intelligence research [research frontier]. *IEEE Computational Intelligence Magazine, 5*(4), 13–18.

4. Christopher, O., (2014). *Conv Nets: A Modular Perspective*. https://colah.github.io/posts/2014-07-Conv-Nets-Modular/ (accessed on 18 December 2020).

5. Bouvrie, J., (2006). *One Introduction Notes on Convolutional Neural Networks*. Cogprints.

6. LeCun, Y., Bengio, Y., & Hinton, G., (2015). Deep learning. *Nature, 521*(7553), 436–444.

7. Lee, C. Y., Gallagher, P. W., & Tu, Z., (2016). Generalizing pooling functions in convolutional neural networks: Mixed, gated, and tree. In: *Artificial Intelligence and Statistics* (pp. 464–472).

8. Karpathy, A., (2017). *CS231n: Convolutional Neural Networks for Visual Recognition*. Stanford University. Available at: http://cs231n.stanford.edu/ (accessed on 18 December 2020).

9. Sentdex, (2016). *Full Classification Example with ConvNet*. Retrieved from: http://cogprints.org/5869/1/cnn_tutorial.pdf (accessed on 18 December 2020).

10. Simon, H. N., (2020). *Breast Cancer Detection Using Convolutional Neural Networks*. Accepted as a workshop paper at AI4AH, ICLR.

11. Ben-Ari, R., Akselrod-Ballin, A., Leonid, K., & Sharbell, H., (2017). Domain-specific convolutional neural nets for detection of architectural distortion in mammograms. In: *2017 IEEE 14th International Symposium on Biomedical Imaging (ISBI 2017)* (pp. 552–556). IEEE.

12. Freddie, B., Jacques, F., Isabelle, S., Rebecca, L. S., Lindsey, A. T., & Ahmedin, J., (2018). Global cancer statistics 2018: Globocan estimates of incidence and mortality worldwide for 36 cancers in 185 countries. *CA: A Cancer Journal for Clinicians, 68*(6), 394–424.

13. Ferlay, J., Soerjomataram, I., Ervik, M., Dikshit, R., Eser, A., Mathers, C., Rebelo, M., et al., (2015). *GLOBOCAN 2012 v1.1, Cancer Incidence, and Mortality Worldwide: IARC Cancer Base No. 11*.

14. Breast Cancer (2020). *Mammography: Benefits, Risks, What You Need to Know*. Breastcancer.org Susan Greenstein Orel, M.D Available at: http://www.breastcancer.org/symptoms/testing/types/mammograms/benefits_risks (accessed on 18 December 2020).

15. Bloom, H. J., & Richardson, W. W., (1957). Histological grading and prognosis in breast cancer: A study of 1409 cases of which 359 have been followed for 15 years. *Br. J. Cancer, 11*(3), 359–377.

16. Elston, C. W., & Ellis, I. O., (1991). Pathological prognostic factors in breast cancer. The value of histological grade in breast cancer: Experience from a large study with long-term follow-up. *Histopathology, 19*(5), 403–410.

17. Haibo, W., Cruz-Roa, A., & Ajay, B., (2014). Mitosis detection in breast cancer pathology images by combining handcrafted and convolutional neural network features. *Journal of Medical Imaging, 1*(3), 034003.

18. Irshad, H., (2013). Automated mitosis detection in histopathology using morphological and multi-channel statistics features. *J. Pathol. Inf., 4*(1), 10–15.

19. Sommer, C., et al., (2012). Learning-based mitotic cell detection in histopathological images. In: *Int. Conf. on Pattern Recognition (ICPR)* (pp. 2306–2309). IEEE, Tsukuba, Japan.

20. Irshad, H., et al., (2013). Automated mitosis detection using texture, SIFT features and HMAX biologically inspired approach. *J. Pathol. Inf., 4*(2), 12–18.

21. WHO. (2020). *The World Health Report: World Health Organization* (WHO). https://www.who.int/whr/previous/en/ (accessed on 18 December 2020).

22. Santosh, K. C., (2020). AI-driven tools for coronavirus outbreak: Need of active learning and cross-population train/test models on multitudinal/multimodal data. *Journal of Medical Systems.*

23. Sina, F. A., Amir, M., Pedram, G., Filip, F., Varkonyi-Koczy, A. R., Uwe, R., Timon, R., & Peter, M. A., (2020). COVID-19 outbreak prediction with machine learning. *Journal of Mathematics*, MDPI.

24. Martínez-Álvarez, Asencio-Cortés, G., Torres, J. F., Gutiérrez-Avilés, D., Melgar-García, L., Pérez-Chacón, R., Rubio-Escudero, C., et al., (2020). *Coronavirus Optimization Algorithm: A Bioinspired Metaheuristic Based on the COVID-19 Propagation Model.* Cornell University.

25. Kamel, B. M. N., & Geraghty, E. M., (2020). *Geographical Tracking and Mapping of Coronavirus Disease COVID-19/Severe Acute Respiratory Syndrome Coronavirus 2 (SARS-CoV-2) Epidemic and Associated Events around the World: How 21ˢᵗ Century GIS Technologies are Supporting the Global Fight Against Outbreaks and Epidemics.*

26. Ali, N., Ceren, K., & Ziynet, P., (2020). *Automatic Detection of Coronavirus Disease (COVID-19) Using X-Ray Images and Deep Convolutional Neural Networks.* Cornell University.

27. Lilian, W. (2017). *GitHub.* https://lilianweng.github.io/lil-log/2017/12/31/object-recognition-for-dummies-part-3.html#model-workflow (accessed on 18 December 2020). Lil log.

28. Russakovsky, O., Deng, J., Su, H., Krause, J., Satheesh, S., Ma, S., Huang, Z., et al., (2015). Image net large scale visual recognition challenge. *International Journal of Computer Vision, 115*, 211–252.

29. Henschke, I., Yankelevitz, D. F., Mirtcheva, R., McGuinness, G., McCauley, D., & Miettinen, O. S., (2002). Ct screening for lung cancer: Frequency and significance of part-solid and nonsolid nodules. *American Journal of Roentgenology, 178*(5), 1053–1057.

30. Uijlings, J., Van, D. S. K., Gevers, T., & Smeulders, A., (2013). Selective search for object recognition. *IJCV.*

31. Wei, T., & Dongsheng, Z., (2019). A two-stage approach for automatic liver segmentation with faster R-CNN and deep lab. *Neural Computing and Applications.*

CHAPTER 3

TECHNICAL ASSESSMENT OF VARIOUS IMAGE STITCHING TECHNIQUES: A DEEP LEARNING APPROACH

M. ANLY ANTONY,[1] R. SATHEESH KUMAR,[2] G. R. GNANA KING,[3] V. YUVARAJ,[4] and CHIDAMBARANATHAN[5]

[1]*Assistant Professor in CSE, Sahrdaya College of Engineering and Technology, Kodakara, Thrissur, Kerala, India,*
E-mail: anlyantony@sahrdaya.ac.in

[2]*Associate Professor in CSE, Sahrdaya College of Engineering and Technology, Kodakara, Thrissur, Kerala, India,*
E-mail: satheeshkumar@sahrdaya.ac.in

[3]*Associate Professor in ECE, Sahrdaya College of Engineering and Technology, Kodakara, Thrissur, Kerala, India,*
E-mail: gnanaking@sahrdaya.ac.in

[4]*Associate Professor in BM, Sahrdaya College of Engineering and Technology, Kodakara, Thrissur, Kerala, India,*
E-mail: yuvarajv@sahrdaya.ac.in

[5]*Associate Professor in MCA, St. Xavier's College, Tirunelveli, Tamil Nadu, India*

ABSTRACT

This chapter presents various image-stitching techniques based on different feature extraction methods. Image stitching can be regarded as a process of assembling more than one image of the same scene having an overlapping area in between them into a single high-resolution image. Direct as well as feature-based methods, are two broad categories of image stitching techniques. Direct methods are more time consuming as compared to feature-based techniques since it needs to compare each and every pixel intensities with each other. Image stitching on the overlapping region is divided into the following steps: first features are detected and described using any kind of feature extraction; secondly, find matching pairs followed by removing mismatches by RANSAC (random sample consensus), then estimate the homography matrix; and finally, blend the overlapping areas. The authors introduce a complete framework for feature-based automatic image stitching. Stitching images where overlapping regions are missing is also a very hot topic of research. It follows image extrapolation, alignment, and in painting.

3.1   INTRODUCTION

Image stitching technology is a very decisive research area in the field of image processing. Due to the accelerated deployment of the internet, multimedia, and graphics technology, people came into contact with an extensive number of images and video information. Image processing techniques need a large amount of detail within the image itself. However, it may not be easy to collect and evaluate details from a single image. There comes the need of multiple images and creation of composite image. Image stitching can be regarded as a process of assembling more than one image of the same scene having overlapping area in between them into a single image with high resolution. This process can also be termed as photo stitching or image mosaicing. It helps to create panorama from multiple input images of the same scenes. Panoramas are expressions of virtual reality with high resolution [29]. It is considered as an elementary way to replicate a reality scene. Panoramas are a popular image based rendering technique. In order to develop a panorama, a mosaic of images of different regions of the same scene is needed. This can be achieved by aligning all images into the same coordinate system and then blending the overlapping regions smoothly [1].

Image stitching system can even be implemented in a mobile phone, so that the users can collect a sequence of images in a wide range of scene and create a panoramic image dynamically and share it with others.

Image stitching/mosaicing was mainly introduced to defeat the problem of the narrow field of view of compact cameras and cameras on mobile devices. A camera can take pictures within the scope of its limited field of view only; it cannot take a high-resolution image with all sort of details fitted in just one single frame. Normally compact cameras are having a field of view around 50×35°, whereas the human visual system has a much larger field of view around 200×135 degrees [2]. A panoramic mosaic will have a field of view much higher than the human visual system. It can be extended up to 360×180 degrees.

The major application of image stitching system lies in the field of video matting, video stabilization, satellite imaging, and medical applications. Image stitching techniques can even be extended to video compression and summarization. Even multi-image stitching can be generalized to video stitching. Image stitching has got wide applications in the medical field for diagnosing cardiac, retinal, pelvic disorders [13]. Microscopy and radiology are relevant areas where image stitching can be applied if the specimen is larger than a line of vision of a camera or microscope. It may also be used in satellite imaging for digital map creation. It can also be used for object removal applications. It needs operations to remove foreground and then estimating the background in order to fill the gaps.

Image stitching approaches are predominantly classified into two categories: direct and feature-based techniques. All the pixel intensities are compared with each other in direct methods, but the latter compare only the extracted features from the image. Therefore, it is fast as compared to direct methods. Direct methods need high quality images and it focus to reduce the total sum of absolute differences between each pixels in the overlapping region. A correlation-based scheme that operates on direct pixel is used in 2011 [10] for image stitching on frequency domain. It makes use of fast Fourier transformations instead of direct Fourier transformations for estimating parameters of motion model. Image stitching can also be performed using FFT phase correlation. Pixel mapping methods are employed for stitching the phase-correlated parameters. Direct techniques do not remain invariant to rotation and scaling [19]. Because of these reasons, feature based techniques are much more convenient comparing to direct methods.

Features are special points within the image that are distinct, trackable, and easily comparable too. These invariant points are extracted and used to define the neighborhood for feature description. Then neighborhood points are used to perform feature matching. It forms the feature descriptor of the feature to be detected. The most commonly used feature-based methods are SIFT [15], SURF [4], ORB [21], Harris corner features [12], multi-scale oriented patches (MSOP) [9], etc.

Image registration and blending are two essential parts of the image stitching process. Image registration is considered as the core part of image mosaicking. The task of image registration is the establishment of matching between the images. It leads to aligning images having different viewpoints to the same coordinate system [30]. It involves feature detection, feature matching, and transform model estimation. Image blending aims at making the transition between images seamless [18]. By making the blending process artifacts free, it finally results in an attractive, visually plausible panorama. Smoother transition makes the entire image stitching process much more efficient.

The reminding portion of the chapter is well organized as follows. In the next section of the chapter, researchers recapitulate the related works done in this area. The proposed mechanism is detaily narrated following background study, in which a detailed illustration of the feature selection method adopted for the study is explained. Then discuss the details of alignment and stitching on non-overlapping images. The next section describes the experimental setup and results. Future research problems will be discussed next. Final section presents the conclusion of the study.

3.2 BACKGROUND

Direct, as well as feature-based methods are two broad categories of image stitching. In the direct method, Richard and Heung-Yeung [23] put forward a method in which panorama is created through a set of transformations. Since it recovers 3D rotations, they are fast and robust. The frequency-domain can be used to directly stitch the images. Durga Patidar et al. used a correlation-based scheme for image stitching, which makes use of FFT (fast Fourier transform) for image alignment in order to calculate transformation parameters [10]. In this work, the authors proposed FFT phase correlation to perform image registration and make use of these parameters for pixel mapping.

Different types of methods can be employed for extracting features from images. As a pioneer work in feature-based image retrieval, Schmidt and Mohr [22] uses a rotation-invariant descriptor by extending Harris corner by gaussian derivation. Later Mattew Brown and David Lowe [8, 15, 29] proposed SIFT (scale-invariant feature transform) features for feature extraction. It provides scale-invariant features by approximation of Laplacian of Gaussian using difference of Gaussian (DOG). The main benefit of SIFT features are robustness and maximum feature extraction. SIFT finally makes a 128-dimensional feature vector. MOPS (multi-scale oriented patches) features are positioned at Harris corners and conformed using a blurred local gradient [9, 12]. This makes it rotation invariant. Low frequency sampling makes feature insensitive to noise. Herbert Bay et al. uses a Speeded up robust features for feature extraction (SURF), which is a speeded-up version of SIFT [4, 13]. It approximates Laplacian of Gaussian with box filter. Integral images are used for performing convolution with a box filter, which reduces calculation time. SURF uses determinant of Hessian matrix for both scale and location invariance. It also reduces the feature space to 64 dimensions. Another improved feature extraction technique is through the use of ORB (Oriented FAST and Rotated BRIEF) features [21].

It is a consolidation of FAST (features from accelerated segment test) keypoint detector and BRIEF (binary robust independent elementary features) descriptors. ORB feature extraction saves computation time and completely free of any licensing restrictions, which is mandatory for SIFT and SURF. ORB is almost two times faster in matching speed than SIFT if images do not have much scale invariance. It performs poor with scale changes. Yanyan Qin et al. proposed another method for feature point matching via improved ORB [27]. It mainly focuses on the weakness of ORB with scale variance. It combines ORB with SIFT so that scale-invariant feature points are detected and which leads to scale-invariant feature descriptor. SIFT is used for feature detection and ORB descriptors make rotation invariant feature descriptors.

As mentioned earlier, image stitching approaches belong to image registration and blending. Image registration has been well explained in Ref. [30], which involve registration model selection. Then it uses the Levenberg-Marquardt algorithm for refining the motion parameter. Zhong, Zeng, and Xie [29] presents a survey on existing image registration techniques which make use of feature detection, feature matching, estimating transform model, and concluding with transformation and image resampling. It proposes various method like feature-based, area, and correlation-based techniques to carry out image registration. Image blending is performed in 2005 [18] using a gradient-based method. It is useful in the presence of sharp intensity changes. It gradually blends input images to a composite image. Gracias et al. [11] Present a very fast image blending technique based on graph-cuts and watershed algorithm. It offers the optimal solution for the intersecting region. It can create large mosaics without user interaction.

3.3 IMAGE STITCHING

This section will cover the details of the image stitching process. It includes components of image stitching, important feature extraction methods, Stitching on overlapping images and its phases and finally stitching non-overlapping images.

3.3.1 COMPONENTS OF IMAGE STITCHING

The main units of any image stitching systems are camera calibration, image registration, and image blending. Calibration reduces the differences in ideal lens and camera lens combination used to acquire images [2]. Image registration is the heart of stitching procedure. Image registration involves matching features or finding relationships in a set of images [3]. It precisely related to the success rate (accuracy) or pace of the image mosaicing process. It is the process of inculcating correlation between images of the same scene and aligning them properly. Image registration follows finding out transformation matrix for estimating the model parameters. RANSAC (random sample consensus) [14] can be used for this purpose. It is a non-deterministic algorithm which never promises to return good results always. The algorithm prunes the outliers after feature matching in order to refine the matched feature set. It starts by randomly taking the feature points and estimating the model. It counts the number of remaining features which fit to the model as inliers and others as outliers. It keeps the model with the largest set of inliers as the transformation of homography matrix.

Image blending aims at making the transition from one image to another much more smoother in order to make seamless stitching. There exist two common methods for performing blending. They are feathering or alpha blending and Gaussian pyramid blending. In alpha blending, pixel values in the blended region are obtained by taking the weighted average of pixel intensities from images to be blended [16]. Gaussian pyramid blending [13] represents an image at different resolution level using image pyramid. It merges images at different frequencies (Figure 3.1).

Boundary blurring has an inverse proportion with frequency, i.e., more blurring occurs with lower frequency band. Gaussian pyramid make pixels near to the boundary pixels more blurred, while preserving pixels as it move far away from the boundary of the stitching line.

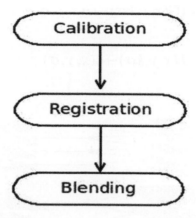

FIGURE 3.1 Components of image stitching.

3.3.2 FEATURE EXTRACTION TECHNIQUES

The researchers have proposed various techniques and algorithms for feature extraction such as SIFT, SURF, ORB methods for feature extraction. Each one of them is described in detail in subsections.

3.3.2.1 SCALE INVARIANT FEATURE TRANSFORM (SIFT)

The exertion of the SIFT algorithm begins with scale-space extrema detection to identify the location of key points, and then feature descriptors are produced using keypoints of neighborhood gradient [7, 8, 15]. It is categorized into four steps:

1. **Scale Space Extrema Detection:**
   i. **The Establishment of Difference of Gaussian (DOG) Pyramid:** Scale-space is produced from convolution of input image I(x,y) with Gaussian kernel.

$$L(x, y, \sigma) = G(x, y, \sigma) * I(x, y) \tag{1}$$

where; L: scale-space; I(x, y): image pixel; and σ: scale factor.

Gaussian pyramid has O orders, and each order includes S levels of scale image (Figure 3.2). DOG pyramid is computed from the difference of two nearby scale-space functions:

$$D(x, y, \sigma) = (G(x, y, k\sigma) - G(x, y, \sigma)) * I(x, y) \tag{2}$$

$$= L(x, y, k\sigma) - L(x, y, \sigma) \tag{3}$$

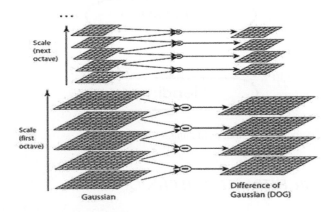

FIGURE 3.2   Difference of Gaussian pyramid.

ii. **Scale-Space Extrema Detection:** Now, to detect the local maxima and minima, the point (marked with X) is compared with its eight neighbors in the current input image and nine neighbors of the same image in the scale below and above it. The point is selected only if it is greater or lesser comparing to all its neighbors (Figure 3.3).

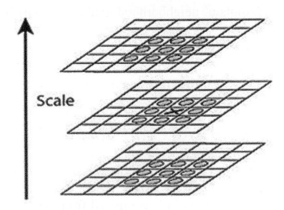

FIGURE 3.3   Scale-space extrema detection.

2. **Keypoint Localization:** Now determine the accurate location and scale of keypoints. The key points with low contrast or that are poorly localized are avoided. Thus the stability of matching is enhanced and noise immunity is improved.

3. **Orientation Assignment:** Each feature point is given a specific orientation parameter. Modulus and direction are defined by:

$$m(x, y) = \sqrt{L_1^2 + L_2^2} \tag{4}$$

$$\theta(x, y) = \arctan(L_2 / L_1) \tag{5}$$

$$L_1 = L(x+1, y, \sigma) - L(x-1, y, \sigma) \tag{6}$$

$$L_2 = L(x, y+1, \sigma) - L(x, y-1, \sigma) \tag{7}$$

where; m(x, y): modulus; and θ(x, y): direction.

4. **Generating SIFT Descriptor:** The contents over 4×4 sub-regions are summarized with orientation histograms with eight directions and the cumulative value for each gradient direction from the seed point. To improve the stability, each feature point is estimated by using a 4×4 point array having eight directional information. This leads to the formation of a 128-dimensional SIFT feature vector (Figure 3.4).

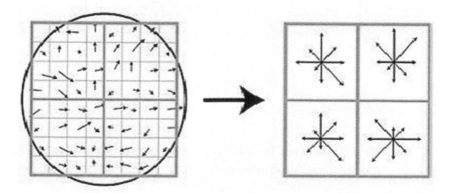

FIGURE 3.4 SIFT descriptor.

3.3.2.2   SPEEDED UP ROBUST FEATURES (SURF)

It makes use of Hessian matrix-based approximations making use of integral images [4, 13]. Descriptors describe the distribution of intensity content within the interest point neighborhood. Authors present a new indexing method based on the sign of Laplacian, which increases the matching speed and robustness.

1. **Interest Point Detection:** It lends to the use of integral images and computation time is scaled down drastically. It involves the following:

    i. **Integral Images:** Integral image allows fast computation of box type convolution filters. $I_\Sigma\,(x)$ at $X = (x, y)$ denotes the sum of all pixel in the input image I within the rectangular region formed by origin and X.

$$I_\Sigma\,(x) = \sum_{i=0}^{i \leq x} \sum_{j=0}^{j \leq x} I(i, j) \tag{8}$$

It will produce the result by only three additions, which compute the sum of intensity value over any rectangular area, and the calculation time never depend on its size (Figure 3.5).

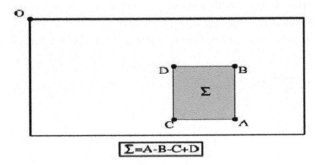

FIGURE 3.5   Integral image.

    ii. **Hessian Matrix Based Interest Points:** Hessian matrices are having good performance inaccuracy so that blob-like structures can be identified at locations where determinant is maximum. Given a point $X = (x, y)$ in an image I, the Hessian Matrix $H(x, \sigma)$ in X at scale σ is defined as follows:

$$H(X,\sigma) = \begin{pmatrix} L_{xx}(X,\sigma) & L_{xy}(X,\sigma) \\ L_{xy}(X,\sigma) & L_{yy}(X,\sigma) \end{pmatrix} \qquad (9)$$

Here; $L_{xx}(X,\sigma)$-convolution of the Gaussian second-order derivative $\frac{d^2}{dx^2}g(\sigma)$ with the image. Gaussians are best for scale-space but discretized and cropped, and this leads to loss of reproducibility around odd multiples of $\pi/4$. The detectors still perform well due to the square shape of the filter, and the tiny decrease in performance does not overweigh the advantage of fast convolutions. These second-order Gaussian approximations can be estimated at very low computation cost by using integral image. The calculation time is self-reliant of the filter size. The 9×9 filter is an approximation of a Gaussian with σ = 1.2. It shows the lowest scale. The rectangular regions are assigned with a simple weight. This result in:

$$\det(H_{approx}) = D_{xx}D_{yy} - (wD_{xy})^2 \qquad (10)$$

Here w, relative weight is used to balance Hessian's determinant expression and is given by:

$$w = \frac{|L_{xy}(1.2)|_F |D_{yy}(9)|_F}{|L_{yy}(1.2)|_F |D_{xy}(9)|_F} \qquad (11)$$

where; $|x|_F$ is the Frobenius norm. The filter responses are normalized with respect to its size and maintain a constant Frobenius norm. These are stored in a blob response map over different scales.

iii. **Scale Space Representation:** Interest points at different scale found due to the search for correspondence requires comparison of image at different scale. Scale spaces are constructed usually using an image pyramid. Images are smoothened multiple times with a Gaussian and then subsampled in order to develop a higher level of the pyramid. The scale space is evaluated by scaling-up the filter size instead of iteratively reducing the size of the image. The output of 9×9 filters will form the initial scale layer, and the following layers are found by filtering the resulting image with increasing the mask size gradually. Here the main motivation is its computational efficiency. Since there is no downsampling of the image, there is no aliasing.

The scale space is decomposed to pieces called octaves. It is a sequence of filter response maps obtained by input image convolution with a filter which increases its own size. Each octave will be broken down to a consistent number of scale levels. Minimum scale difference between two subsequent scales relies on the length l_0 of the negative or positive lobes of the partial second-order derivative. l_0 is set to one-third of filter size length. Then increase the resulting size by a 2 pixels (minimum) to keep the size uneven and to ensure the presence of central pixel.

The construction of scale-space start with (9×9) filters then with filter of size 15×15, 21×21, and 27×27. For each of the new octave, the filter size is increased by a factor of two. The filter sizes of the second octave are 15, 27, 39, and 51. The third octave with filter size 27, 51, 75, 99, and so on. The large scale changes, makes the sampling of scales difficult. Therefore, it establishes a scale-space with a finer scale sampling. It first magnifies the size of the image by 2 and then create the first octave by using a filter of size 15. Additional filter size 21, 27, 33, and 39. An increase of filter size by 12 pixels is applied to the second octave. Now, scale changes between the first two filters are 1.4 (21/15).

    iv.   **Interest Point Localization:** For localizing interest points, a non-maximum suppression is performed in a neighborhood of filter size 3×3. The maxima of determinant of the hessian matrix are interpolated, and it is important because of the huge deviation in scale between the first layers of each octave.

2.   **Interest Point Description and Matching:** Descriptor is used to show the distribution of intensity content within interest point neighborhood. It consists of three steps.

First, calculate the Haar wavelet response in x and y direction around the circular neighborhood centered at the interest point. It tries to step down the time for feature computation and matching. An indexing step relying on the sign of the Laplacian is used as the descriptor which magnifies the robustness as well as matching speed. The first step consisting of fixing an orientation from a circular region around the interest point. Following a square region is constructed, followed by extraction of SURF features. Finally, features are matched. The detailed explanation is given below:

    i.   **Orientation Assignment:** The procedure starts with the calculation of the Haar wavelet responses in x and y direction within the circular neighborhood of radius 6s where s is the scale at which interest point was detected. It requires six operations to compute the response in both x and y direction at any scale. Once wavelet responses are calculated and weighted with a Gaussian ($\sigma = 2s$), the responses are plotted as points in space. The dominant orientation is determined by finding the total sum of all responses bounded by a sliding window of size $\Pi/3$ as in Figure 3.6. Then summation on horizontal and vertical responses is calculated to find the local orientation vector. The longest vector shows the orientation of the interest point. The size of the sliding window need to be chosen carefully; otherwise results in misorientation. This is done for rotation invariance.

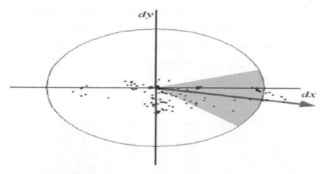

FIGURE 3.6 Orientation assignment.

ii. **Descriptor Based on Sum of Haar Wavelet Responses:** First step consisting of constructing a square region centered on interest point. It is then oriented along the selected orientation. The window size is 20s. The region is splitted up into smaller 4×4 square sub-regions (Figure 3.7).

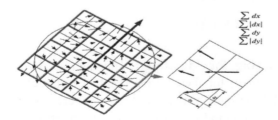

FIGURE 3.7 SURF descriptor.

For each sub-region, the Haar wavelet responses are calculated along x ^dx and y as ^dy direction followed by summation on wavelet responses over each sub-region and form a set of entries in feature vector. For each sub-region has a 4-dimensional descriptor vector **v** for its underlying intensity structure.

$$v = (\sum d_x, \sum d_y, \sum |d_x|, \sum |d_y|) \tag{12}$$

Coupling this for all 4×4 sub-region, a feature vector of length 64 is obtained. Figure 3.8 shows the properties of the descriptor. For homogeneous regions, the values of feature vector parameters are relatively low. If there exist frequencies in x-direction, then the value $\sum |d_x|$ is high. If intensity is gradually increasing in x-direction both $\sum d_x$ and $\sum |d_x|$ will be high. SURF is keen on the spatial distribution of gradient information, and it beat SIFT in many cases. This is because SURF integrates the gradient information, whereas SIFT depends on the orientation of individual gradients. So, SURF is less affected by noise.

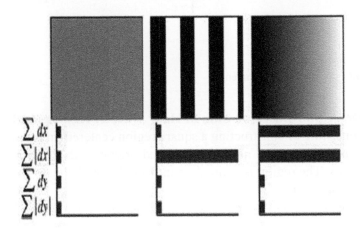

FIGURE 3.8 Properties of descriptor.

iii. **Fast Indexing for Matching:** The sign of Laplacian (trace of Hessian matrix) for the interest point is considered. It identifies a bright blob on dark background from reverse situation. Advantage of using this method is that no extra computational cost is required. In the matching step, a comparison of features, if they have same type of contrast is shown in Figure 3.9. Hence minimum amount of data allows for faster matching without reducing the descriptor's performance.

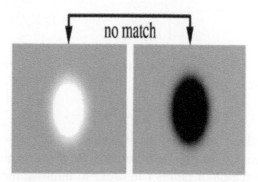

FIGURE 3.9 Comparison of features.

3.3.2.3 ORIENTED FAST AND ROTATED BRIEF (ORB)

ORB methods have been widely used because of their computational properties. It is having the additional advantage of good performance and reduced cost [21]. It is also less affected by noise. It is based on the FAST key point detector and BRIEF descriptor. So let us first discuss about oriented FAST.

1. **FAST-FAST Keyboard Orientation:** FAST features lack an orientation component. So, add an efficiently computed orientation with it.
   i. **FAST Detector:** FAST consider the intensity threshold between the center pixel and the pixels in a circular region about its center with a constant radius as the only parameters. It does not produce cornerness and FAST has large responses along edges, Harris corner measure is added for ordering the key points. FAST never produce multiple-scale features, so compute FAST features at each level of an image pyramid.
   ii. **Orientation by Intensity Centroid:** It uses an effective measure of corner orientation as intensity centroid. It assumes that intensity is offset from its corner, and this vector is used to impute an orientation. Moment is defined as follows:

$$m_{pq} = \sum_{x,y} x^p y^q I(x, y) \tag{13}$$

Centroid is defined as follows:

$$C = \left(\frac{m_{10}}{m_{00}}, \frac{m_{01}}{m_{00}} \right) \tag{14}$$

A vector $\overrightarrow{OC}$ is constructed from the center to the centroid. Then orientation is given by:

$$\theta = a\tan 2(m_{01}, m_{10}) \tag{15}$$

where; $a\tan_2$-quadrant aware version of arctan.

It is used to ensure that moments are computed with x and y that remain within a circular region of radius r in order to improve the rotation invariance. As $|C|$ reaches 0, the measure become unstable but it occurs rarely.

2. **rBRIEF-Rotation Aware BRIEF:** ORB uses a steered BRIEF descriptor and then put forward a learning step to obtain minimally correlated binary tests that leads to better descriptor rBRIEF.

   i. **Efficient Rotation of the BRIEF Operator:** BRIEF descriptor is a form of bit string description about an image patch developed from a set of binary tests on intensity values. A binary test τ on image patch p is defined by:

$$\tau(p; x, y) = \begin{cases} 1 \ if \ p(x) < p(y) \\ 0 \ if \ p(x) \ge p(y) \end{cases} \tag{16}$$

where; p(x): intensity of a patch p centered at a point x.

The feature is defined as a vector of n binary tests:

$$f_n(p) = \sum_{1 \le t \le n} 2^{t-1} \tau(p; x, y) \tag{17}$$

➤ **Steered BRIEF:** In order to steer BRIEF orientation of keypoints are taken. A 2×n matrix will be taken for any feature set consisting of at least n binary tests at (x_i, y_i),

$$S = \begin{pmatrix} x_1, x_2, ..., x_n \\ y_1, y_2, ..., y_n \end{pmatrix} \tag{18}$$

Now the steered version S_θ of S:

$$S_\theta = R_\theta S \tag{19}$$

The steered BRIEF operator becomes:

$$g_n(p,\theta) = f_n(p) \,|\, (x_i, y_i) \in S_\theta \tag{20}$$

As the keypoint orientation θ does not change across views, S_θ will be used to estimate its descriptor.

ii. **Variance and Correlation:** BRIEF has large variance and a mean near 0.5 for each bit feature. Once BRIEF is oriented about the keypoint direction in order to produce a steered BRIEF, the means are actually shifted to a pattern which is more distributed. Very large variance makes a feature more distinctive. Another property is to have tests uncorrelated. Both BRIEF and steered BRIEF shows large initial eigenvalues denoting correlation between the binary tests. Steered BRIEF has significantly lower variance, but it is not discriminative since eigenvalues are lower.

iii. **Learning Good Binary Features:** Choose a good subset of binary tests that reduce correlation and loss of variance. From the existing set, take a large one and identify 256 features having high variance and are uncorrelated using some dimensionality reduction method. We list all possible binary tests drawn from 31×31 pixel patch, and each test is a pair of 5×5 sub-image window. If the width of our patch $w_p = 31$ and width of test sub-window as $w_t = 5$, then we have, $N = (w_p - w_t)^2$ where w_p is the width of patch and w_t is the width of test sub-window and we select pairs of two elements from these so $C(N,2)$ binary test will be chosen. $M = 205{,}590$ by eliminating overlapping tests. The algorithm is as follows:

➢ **Algorithm**
- **Step 1:** Execute each test against all training patches.
- **Step 2:** Make a sequence by arranging the tests by their distance from a mean of 0.5, which leads to the formation of the vector.
- **Step 3:** Greedy search:
  - Take the first test and add it into the result vector R and then remove it from T, the Test set.

> o   Consider the next test from T, and compare it with all tests in R for finding the absolute correlation. If its absolute correlation is higher than an estimated threshold, omit it, else attach it to R.
>
> o   Repeat this procedure for all the 256 tests in R. If the count is less than 256, increase the threshold and try again.

The result generates rBRIEF. This algorithm is categorized as a greedy search which is applicable to a set of uncorrelated tests with means near 0.5.

iv.   Scalable Matching of Binary Features: Locality sensitive hashing (LSH) [28] is used for nearest neighbor search. In LSH, hash tables are storing the points and it is hashed to different bucket slot. For any query descriptor, its matching buckets will be found and returned. The elements are compared using a brute force matching. For binary features, the hash function is a subset of signature bits and the same bucket contain descriptors of the same signature. Hamming distance is used as a distance metric.

3.3.3   CASE STUDY

Image stitching consists of several steps as feature detection, feature matching, feature refining through RANSAC, homography matrix estimation, image alignment, and image blending, as shown in Figure 3.10.

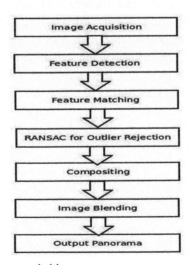

FIGURE 3.10   Steps in image stitching.

- **Step 1: Image Acquisition:** It is the first stage of any image stitching system. It is the process of collecting the images from its sources.
- **Step 2: Feature Detection:** Special points like corners, blobs are detected within the image and are used to find relationship or correspondences between images. Once features are detected, they are described using their neighborhood.
- **Step 3: Feature Matching:** Computed features can be matched using a BF matcher or FLANN matcher for finding the overlapping region.
- **Step 4: Mismatch Removal:** It aims at removing the outliers from matched features for refining the final set of matched features.
- **Step 5: Homography Matrix:** RANSAC is used to estimate parameters of the transformation matrix since it is essential to convert one image to another image's coordinate system.
- **Step 6: Image Alignment:** It is the process of finding the projection model and warping the images into the same coordinate system as that of the reference image.
- **Step 7: Image Blending:** It can be regarded as the process of formulating the transition between images much more smoother using alpha blending or Gaussian pyramid blending.

3.3.3.1 FEATURE DETECTION

It would be beneficial to find some special points within the image and then perform local analysis on these ones only instead of taking the whole image. Features are elements in the input images to be matched. Transformative relations can be established from the common features in the overlapping area. There exist a number of techniques for describing the local image regions. In the proposed method, we use ORB feature detection [21]. It is a consolidation of FAST detector and BRIEF descriptor. FAST detector analyzes pixels on a circle of fixed radius around the feature point which is to be taken as a key point. A point will be considered as a corner only if at least n pixels on a circle of fixed radius that cover 16 pixels on its boundary around the feature point are all found as brighter or reverse. FAST keypoint detection proceeds as follows. First it takes a circle of 16 pixels around the candidate pixel as shown in Figure 3.11 and is selected as a feature only when a set of n continuous pixels are all found as brighter than candidate pixel + T or darker than <=T, where is the pixel intensity and T is the threshold. This process can be accelerated by applying a high-speed criterion on it. In that case, first check for $^I1$, $^I5$, $^I9$, $^I13$ pixels on the major and minor axis of the circle around candidate pixel. Then check whether at least three of them are above or below + T. If so, then only check about all 16 pixels to ensure whether 12 continuous pixels satisfy the criterion, otherwise simply discard the pixel which is not a corner.

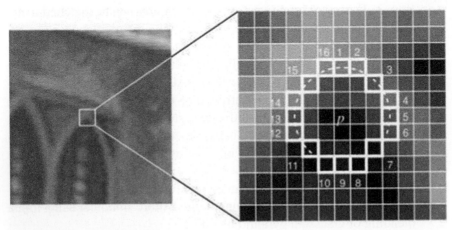

FIGURE 3.11   FAST detector.

Now for descriptors, ORB use BRIEF descriptors. BRIEF introduces an approach to find 32-bit binary strings directly. Take first location pairs be (p,q). If Intensity of p,I(p) > Intensity of q, I(q), then result will be 0, else it is taken as 1. This is applied for all the location pairs in order to get a dimensional bitstring. However, BRIEF does not perform well with rotation. So keypoint orientation is used to steer BRIEF. For each feature set of n binary tests at location (,), determine a 2×n matrix, S in which each row contains the x and y coordinates of these pixels. Then making use of the orientation of selected patch as, the rotation matrix of the same is found and then rotate or steer S to find the steered (rotated) version of it as.

3.3.3.2   FEATURE MATCHING

Once the interested features are detected, it is then taken into the next level. In the feature matching step, features in one image are matched with others. The detected feature points are described unambiguously to compute correspondence between the images. For better feature matching, corner like features need to be matched. There are two approaches to perform matching:

Brute Force Matcher and FLANN-based matcher. Here we use FLANN (fast library for approximate nearest neighbors) for feature matching. In FLANN, LSH [28] for nearest neighbor search. A set of hashing functions will be calculated in order to keep the features that are close to each other, will be mapped with the same hash value. For each hash function, retrieve all the feature descriptors assigned to the same hash function. Then we need to compute the distance on this short-list only. Normally, Hamming distance can be used for distance calculation. K-nearest neighbors is returned for each feature points. K-value is taken as two, thereby returning the best two matched features. Out of these two matches, the best one is selected using lowest ratio test [5].

3.3.3.3 RANSAC FOR MISMATCH REMOVAL

It is usual to find correspondences using a weighted sum of squared differences within a small neighborhood around the features. However, these are not robust to geometric distortions. So RANSAC [14] can be used for eliminating the outliers. It is a non-deterministic algorithm. When RANSAC is being used for feature based image matching, we need to find the transform that best translate the first image to the second image. It can be applied to get the homography of each image pair. Four initial feature matches are selected randomly in each iteration of the RANSAC algorithm. It is used to estimate homography parameters. Final iteration will result in a correct homography estimation only if feature matches are real inliers. RANSAC consist of the following steps:

- **Step 1:** Randomly choose s samples.
- **Step 2:** Fit a model (e.g., line) to those samples.
- **Step 3:** Count the number of inliers that approximately fit the model.
- **Step 4:** Repeat N times.
- **Step 5:** Choose the model that has the largest set of inliers.

3.3.3.4 HOMOGRAPHY MATRIX ESTIMATION

A homography matrix is used to calculate the coordinate transformation relation between images. Given point correspondences X_1 and $^NC_2$ then similarity transformation T_1 is given as:

$$T_i = \begin{pmatrix} t_1 & t_2 & t_3 \\ t_4 & t_5 & t_6 \\ t_7 & t_8 & 1 \end{pmatrix} \tag{21}$$

which consists of a translation and scaling in x and y-direction. It will take points X_i to the new set of points $\tilde{X}_i$ which satisfies the following relation:

$$\tilde{X}_i = T_i * X_i$$

where;

$$X_i = (x_i, y_i, 1)^T \text{ and } \tilde{X}_i = (\tilde{x}_i, \tilde{y}_i, 1)^T \tag{22}$$

3.3.3.5   IMAGE ALIGNMENT

This process is done by selecting a final compositing surface for stitched image once images are registered. First step is to decide how to represent the final image [19]. A basic and fundamental approach is to choose any one image as the reference and then to warp all remaining image to the coordinate system of the referred image as per the homography matrix. If the input images are unordered, then an extra effort is taken to decide which image sequence form one or more panoramas.

3.3.3.6   IMAGE BLENDING

Once the input images are mapped to the final composite surface next step is to blend them to create an appealing panorama. There are different blending methods such as feathering [16, 25], gradient-domain and pyramid blending. The approach used here is known as feathering. It is also termed as alpha blending. In feathering, the pixel values in the blending region will be calculated as the weighted average of pixels taken from overlapping area. It is the simplest approach. It works well if the images to be stitched are taken at the same time without much rotation of the camera. If images are properly aligned, then alpha blending is enough to produce excellent result.

3.3.4 STITCHING NON-OVERLAPPING IMAGES

Previous sections provide details about the steps to be followed if input images are having an area of overlap between them. It needs to find features, matching features in order to determine the area of overlap, etc. However, what if a traveler takes multiple photos of some places and when back at home realizes that these photos do not have enough area of overlap to create a panorama. Non-overlapping image stitching provides a solution for this scenario. There is still hope to create a panoramic image using non-overlapping input images [5, 17]. It needs to follow the following subsections.

3.3.4.1 IMAGE EXTRAPOLATION

Since overlapping area are missing in input images, input images are extrapolated beyond their original boundary to create some overlapping area between them. While creating extrapolated area, it is important to infer from the content inside the image. It needs to find out patches within the image itself to complete the unfilled region. The selection of such a patch depends on self-similarity. It can be find out by applying some kind of distance between the patches. Most commonly used distance functions are Euclidean distance, Bhattacharya distance, Hellinger distance, or Chi-square distance. After finding such a most similar patch, the gaps on the extrapolation area are filled with the same. This process will be repeated till extrapolations from different sides are completed. The result need to be repeated on different scales of the Gaussian pyramid for creating a high-resolution image. It will finally lead to a continuous mosaic by filling the non-overlapping regions between the images.

3.3.4.2 IMAGE ALIGNMENT

The extrapolated image needs to be aligned appropriately in order to produce a continuous panorama. Proper alignment searches for a displacement which reduces the sum of pairwise error between the images. Multiple image alignment process proceeds by placing the input images at the same location on top of each other. Then gradually shifting them in each iteration to reduce the alignment error [5]. While calculating the error between unrelated images, it will not result in much change in it. However, it will significantly reduce the error towards shifts on related images.

3.3.4.3   IMAGE INPAINTING

Image inpainting in non-overlapping stitching can be considered as a post-processing activity. It can be done with the help of content-aware Photoshop. Apart from it, two well-known methods exist are Telea and Naiver stokes. Telea [24] uses a Fast marching method for image inpainting. Image inpainting techniques fill the gaps within the images. It proceeds by filling the boundary regions first then moving inwards. This ensures that the region closest to known image portions are filled at the earliest. Naiver stokes [6] uses ideas and theories from fluid dynamics to the inpainting process. It considers image intensity as a stream function. Fluid flow is used here to proliferate the isophote lines from outside region to the portion where inpainting is done. These processes make use of Naiver-Stokes equations of fluid dynamics and thus the name. It carries fluid dynamics thoughts and ideas to computer vision applications.

3.4   SOLUTIONS AND RECOMMENDATIONS

The experiments were conducted on Ubuntu 14.04 platform with the assistance of Intel Core 3 and 4 GB ram. The real-time image stitching system based on feature extraction has implemented a full image stitching system using OpenCV Python library [3, 20]. This work additionally evaluates some generally used techniques of feature extraction such as sift and surf, which are incorporated in image stitching application. Experiments for comparing various feature extraction techniques have also conducted.

3.4.1   DATASET

The proposed methodology was applied on more than ten different groups of images. Two of them are shown in Figures 3.12 and 3.13.

FIGURE 3.12   Dataset 1.

FIGURE 3.13 Dataset 2.

3.4.2 *ORB FEATURE DETECTION AND MATCHING*

Features are extracted from input images using ORB method. Figure 3.14 shows the ORB features found in the input images of Dataset 1. Figure 3.15 shows the ORB features found in the input images of Dataset 2.

FIGURE 3.14 Detected ORB features in dataset 1.

FIGURE 3.15 Detected ORB features in dataset 2.

Figure 3.16 shows the images of Dataset 1 after feature matching. Figure 3.17 shows the input images of Dataset 2 after feature matching.

FIGURE 3.16   Features matched in dataset 1.

FIGURE 3.17   Features matched in dataset 2.

3.4.3   COMPARISON AND EVALUATION

Once input images are registered, feature matched, mismatch refined and blended, then an output panorama is formed. Figure 3.18 shows the output panorama for input images of Dataset 1. Figure 3.19 shows the output panorama for input images of Dataset 2.

FIGURE 3.18 Panorama using dataset 1.

FIGURE 3.19 Panorama using dataset 2.

Tables 3.1 and 3.2 show the comparison between the proposed ORB feature detection method and other well-known methods of feature extraction methods, SIFT and SURF for Dataset 1 and Dataset 2, respectively.

The table shows that the proposed method of image stitching based on ORB feature detection outperforms other methods which follow SIFT and SURF for feature extraction. It also shows that a number of detected features and time taken has an inverse relationship. As the detected features increases, the time taken to perform stitching also increases. ORB takes the least time to stitch input images comparing to all other methods.

TABLE 3.1   Performance Analysis of Feature Detectors for Dataset 1

Feature Detection Method	Number of Features Detected	Filtered Match Count	Inlier Count	Time Taken (s)
ORB	500	81	56	0.989
SURF	1029	233	221	1.506
SIFT	1840	589	468	1.817

TABLE 3.2   Performance Analysis of Feature Detectors for Dataset 2

Feature Detection Method	Number of Features Detected	Filtered Match Count	Inlier Count	Time Taken (s)
ORB	500	70	52	0.270
SURF	1029	227	219	0.519
SIFT	1713	517	502	0.860

3.5   FUTURE RESEARCH SCOPE

Some of the suggestions for future study are listed below:

- Other feature detectors such as F-SIFT, GLOH, LESH, STAR, and LAZY can be used for image stitching.
- A provisional study between different feature detectors can be done to develop solution for affine transformation problem, scale changes, illumination changes, and noisy images.
- Proposed system can also be applied to videos to create dynamic panoramas with some enhancements.
- Various methods can be introduced to remove visible seams in creating photo mosaics.

3.6 CONCLUSION

Image stitching based on feature extraction technique is a challenging research area. It has got many applications such as medical image stitching for diagnosis of diseases, panorama creation, satellite imaging, etc. Since there are a lot of different feature detection algorithms, the selection depends on the problem being solved. This chapter proposes basic ideas of image stitching with overlapping as well as non-overlapping regions and implements a real-time feature extraction based image stitching system using orb as well as compare it with other well-known methods of feature extraction such as sift and surf. SIFT features are rotation and scale-invariant. It is least affected by noise. It can also detect a large number of features. However, execution time is much larger compared to surf and orb. SURF algorithm takes less time compared to SIFT but takes a large amount of time while comparing with ORB. The experimental results have revealed that ORB outperforms other method in terms of rotation and scale invariance and execution time. It needs less memory requirements too. Since stitching images with overlapping regions belongs to an exhausted research area, stitching images missing overlapping area are also introduced in this chapter.

KEYWORDS

- **fast Fourier transform**
- **features from accelerated segment test**
- **image blending**
- **image registration**
- **image stitching/image mosaicking**
- **locality-sensitive hashing**

REFERENCES

1. Adel, E., Elmogy, M., & Elbakry, H., (2014). Image stitching based on feature extraction techniques: A survey. *Proceedings of IEEE Computer Society Conference on Computer Vision, 99*(6), 120–128.
2. Adel, E., Elmogy, M., & Elbakry, H., (2014). Real-time image mosaicing system based on feature extraction techniques. *Proceedings of 9th International Conference on Computer Engineering and Systems (ICCES)*, 339–345.

3. Alexander, M., & Abid, K., (2014). *OpenCV-Python Tutorials Documentation Release 1.*

4. Bay, H., Ess, A., Tuytelaars, T., & Gool, L. V., (2008). SURF: Speeded up robust features. *International Journal of Computer Vision and Image Understanding, 110*(3), 346–359.

5. Benjamin, C., (2014). *Building a mosaic from Non-Overlapping Images.* Carnegie Mellon University, Pittsburgh. PA.

6. Bertalmío, M., Bertozzi, A. L., & Sapiro, G., (2001). Navier-stokes, fluid dynamics, and image and video inpainting. *Proceedings of the 2001 IEEE Computer Society Conference on Computer Vision and Pattern Recognition, 1,* 355–362.

7. Brown, M., & Lowe, D. G., (2003). Recognizing panoramas. *Proceedings of 9th IEEE International Conference on Computer Vision, 2,* 1218–1225.

8. Brown, M., & Lowe, D. G., (2007). Automatic panoramic image stitching using invariant features. *International Journal of Computer Vision, 56*(2), 30–45.

9. Brown, M., Szeliski, R., & Winder, S. A. J., (2005). Multi-image matching using multi-scale oriented patches. *IEEE Computer Society Conference on Computer Vision and Pattern Recognition, 1,* 510–517.

10. Durga, P., & Akshay, J., (2011). Automatic image mosaicing: An approach based on FFT. *International Journal of Scientific Engineering and Technology, 1*(1), 01–04.

11. Gracias, N., Mahoor, M., Negahdaripour, S., & Gleason, A., (2009). Fast image blending using watersheds and graph cuts. *Image and Vision Computing, 27*(5), 597–607.

12. Harris, C., & Stephens, M., (1988). A combined corner and edge detector. *Proceedings of the 4th Alvey Vision Conference,* 147–151.

13. Juan, L., & Gwun, O., (2010). SURF applied in panorama image stitching. *Proceedings of 2nd International Conference on Image Processing Theory Tools and Applications (IPTA),* 495–499.

14. Konstantinos, G., (2010). *Overview of the RANSAC Algorithm.* Retrieved from: http://www.cse.yorku.ca/~kosta/CompVisNotes/ransac.pdf (accessed on 18 December 2020).

15. Lowe, D., (2004). Image features from scale-invariant key points. *International Journal of Computer Vision, 60*(2), 91–110.

16. Morgan, M., & Louis, B., (2013). Weighted blended order-independent transparency. *Journal of Computer Graphics Techniques, 2*(2), 123–141.

17. Poleg, Y., & Peleg, S., (2012). Alignment and mosaicing of non-overlapping images. *Proceedings of International Conference on Computational Photography,* 1–8.

18. Rankov, V., Locke, R. J., Edens, R. J., Barber, P. R., & Vojnovic, B., (2005). An algorithm for image stitching and blending. In: *Biomedical Optics 2005* (pp. 190–199). International Society for Optics and Photonics.

19. Szeliski, R., (2006). *Image Alignment and Stitching: A Tutorial.* USA, Tech. Rep. TR-2004-92.

20. Laganiere, R., (2011). *OpenCV2 Computer Vision Application Programming Cookbook.* Packt Publishing.

21. Rublee, E., Rabaud, V., Konolige, K., & Bradski, G. R., (2011). ORB: An efficient alternative to SIFT or SURF. *IEEE International Conference on Computer Vision (ICCV),* 2564–2571.

22. Schmid, C., & Mohr, R., (2002). Local gray value invariants for image retrieval. *IEEE Transactions on Pattern Analysis and Machine Intelligence, 19*(5), 530–535.

23. Szeliski, R., & Shum, H. Y., (1997). Creating full view panoramic image mosaics and environment maps. In: *Proceedings of Computer Graphics* (pp. 251–258).

24. Telea, A., (2004). An image inpainting technique based on the fast marching method. *Journal of Graphics Tools, 9*(1), 23–34.
25. Vaghela, D., & Naina, K., (2014). A review of image mosaicing techniques. *International Journal on Computer Vision and Pattern Recognition.* ArXive-prints.
26. Vinod, G., & Anita, R., (2013). Image feature point matching via improved ORB. *International Journal of Advanced Research in Electrical, Electronics and Instrumentation Engineering, 2*(12), 6002–6009.
27. Yanyan, Q., Hongke, X., & Huiru, C., (2014). Image Feature point matching via improved ORB. *Proceedings of International Conference on Progress in Informatics and Computing.* 204–208.
28. Zhang, H., & Hossein, S. *Application of Locality Sensitive Hashing to Real-Time Loop Closure Detection.*
29. Zhong, M., Zeng, J., & Xie, X., (2012). Panorama stitching based on SIFT algorithm and Levenberg-Marquardt optimization. *Proceedings of International Conference on Medical Physics and Biomedical Engineering, 33,* 811–818.
30. Zitova, B., & Flusser, J., (2003). Image registration methods: A survey. *Image and Vision Computing, 21*(11), 977–1000.

CHAPTER 4

CCNN: A DEEP LEARNING APPROACH FOR AN ACUTE NEUROCUTANEOUS SYNDROME VIA CLOUD-BASED MRI IMAGES

S. ARUNMOZHI SELVI,[1] T. ANANTH KUMAR,[2] and R. S. RAJESH[1]

[1]*Department of Computer Science and Engineering,*
Manonmaniam Sundaranar University, Tirunelveli, Tamil Nadu, India,
E-mails: heyaruna@gmail.com (S. A. Selvi),
rsrajesh@msuniv.ac.in (R. S. Rajesh)

[2]*Department of Computer Science and Engineering,*
IFET College of Engineering, Gengarampalayam, Tamil Nadu, India,
E-mail: ananth.eec@gmail.com

ABSTRACT

In this trending world of data loads, the medical field has also become a significant boom in all areas. Most of the areas have been refined to cope with the hi-tech world except for some areas where storage has been a great issue. Those areas in the medical field are found out to revamp its operation through the cloud-based network. Radiologists and physicians are one to one entity that depends upon each has many challenges in which neurological disorders took enough area. Several areas in image analysis are yet to be nourished especially segmentation and classification. In the arena of medical image analysis, there is high scope in segmentation and classification. The deep learning (DL)-based approach is proposed for analyzing and finding neurocutaneous syndromes. The new concept of cloud-based convolution neural network (CCNN) has been proposed. The noteworthy of this work is to develop a more accurate result compared to the conventional approach.

This type of network learning is a different approach of certain types of neurocutaneous syndromes. A decision tree classification is an added advantage in CNN which gives solution for many different types of symptoms other than the MRI images. A set of pre-trained GoogLeNet libraries is used for the analysis of MRI images for this work. This idea achieves almost an accuracy of 95%, which really overcomes the current scenarios. In a practical aspect, the clinical findings itself confirm the syndrome, for more accuracy in the result. This study is a useful technique in areas where medical images are limited to the process. This chapter provides you a more innovative cloud convolutional neural network (CNN) for neurocutaneous syndrome in the biomedical field, which is under severe research.

4.1   INTRODUCTION

Conservative machine-learning techniques were able to process the raw data into a meaningful one with limited efforts. Cautious designing and significant space skills are required for constructing an example acknowledgment or AI framework. Discovering the meaningful data from the huge raw data automatically with a set of methodology followed is popularly known as representation learning.

Deep learning (DL) is a piece of AI in computerized reasoning (AI) that has systems equipped for learning unaided or unlabeled information, additionally referred to likewise as profound neural learning or profound neural system. It has been a great success in removing the barrier of dealing unstructured structures in high-dimensional data and is therefore applicable to many domains of real-world application, namely business, medical science, and research.

DL will emerge great heights when the amount of data is increased. The traditional learning algorithms failed to cope with its performance accurately (Figure 4.1).

The deep learning techniques are classified into three categories such as:

1. **Deep Networks for Unsupervised or Generative Learning (GL):** This learning brings the data out of unsupervised networks, especially when a large amount of raw data which is not classified in any order to gain information.
2. **Deep Networks for Supervised Learning:** This learning brings data from neat and classified structured data. They are also called deep discriminative networks.

3. **Hybrid Deep Networks:** This learning brings data from both structured data and unstructured data. It is mainly used for recognizing the
action of human beings using action bank features.

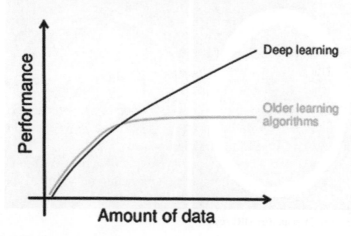

FIGURE 4.1   Why deep learning?

4.2 NEUROLOGICAL DISEASES-NEUROCUTANEOUS SYNDROMES

Neurofibromatosis has a high phenotypic inheritance. So that it can be
perceived in infancy-based on skin abnormalities, especially after puberty
in humans. Inheritance is autosomal prevailing. There are two sorts of
inheritance they are, von Recklinghausen sickness or fringe and peripheral
neurofibromatosis and central neurofibromatosis. Tuberous sclerosis is an
autosomal, overwhelmingly neurocutaneous disorder. Sturge-Weber disorder
is described by facial nevus flammens. These diseases are associated with
CNS pathology. Neurocutaneous syndromes are a group of syndromes characterized by the involvement of the brain and skin. Because both originated
from ectoderm embryologically, they influence the cerebrum, spinal line,
organs, skin, and bones. The ailments are long-lasting conditions that can
make tumors develop in these zones. They can likewise cause different
issues, for example, hearing misfortune, seizures, and formative issues. Each
confusion has various indications.

Figure 4.2(a) shows the multiple foci of periventricular hyper attenuation
consistent with subependymal calcifications of tuberous sclerosis. Figure

4.2(b) is an MRI image that shows the parts of subcortical hyperintensity with tuberous sclerosis.

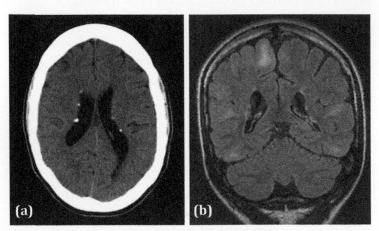

FIGURE 4.2   (a) CT scan; (b) MRI scan.

Neurocutaneous syndromes contain five major syndromes, and they are neurofibromatosis, tuberous sclerosis, Von Hippel Landau disease, Sturge-Weber syndrome (SWS), and Ataxia-Telangiectasia (A-T). Expect SWS, the entirety of the previously mentioned malady, is acquired scatters. The clinical profile of these clutters are varying, transforming from the smooth unsuccessful structures to outrageous conceivably deadly disarranges. The clinical profile of these disorders is various, differing from the gentle unsuccessful structures to the extreme.

4.2.1   NEUROFIBROMATOSIS

Neurofibromatosis is a hereditary illness of the nervous system. It, for the most part, influences how nerve cells frame and develop. It makes tumors develop on nerves. One can get neurofibromatosis from their parents. Inheritance is predominant with autosomal. There are two sorts, such as NF1, which is named von Recklinghausen illness or peripheral NF1, and another type is NF2-central neurofibromatosis. Cancellation or inactivation of the NF gene on chromosome 17 is answerable for NF1, and Gene for NF2 is situated on chromosome 22.

4.2.1.1   NF1

Neurofibromatosis type 1 is an abnormal condition portrayed by changes in skin shading (pigmentation) and the improvement of tumors along with nerves in the skin, cerebrum, and various bits of the body. The signs and reactions of this condition vacillate extensively among impacted people. Beginning in youth, basically, all people with neurofibromatosis type 1 have different cafe Au-Lait spots, which are level fixes on the skin that are darker than the incorporating district. These spots increase in size and number as the individual turns out to be progressively settled. Spots in the underarms and crotch consistently become later in youthfulness.

4.2.1.2   NF2

The closeness of respective sound-related neuroma, one-sided sound-related neuroma nearby a first-degree relative, is with meningioma, schwannoma, or adolescent back subcapsular lenticular obscurity. They are progressively regular in young people: deafness and cerebropontine point disorder (facial palsy + cerebellar ataxia. The highlight of both NF1 and NF2 might be blended. Both may have the endocrinal disorder (MEN): pheochromocytoma, aspiratory hypertension, renal supply route stenosis, glioma of the cerebrum, and sarcoma.

4.2.2   TUBEROUS SCLEROSIS COMPLEX

Tuberous sclerosis is one of the autosomal overwhelmingly neurocutaneous disperse. The features may vary according to age and the condition of the body. The cardinal features can be absorbed via skin wounds, spasm, and mental hindrance. Early skin wounds are hypopigmented, flotsam, and jetsam leaf-shaped.

4.2.3   STURGE-WEBER SYNDROME (SWS)

Sturge-Weber syndrome (SWS) is a neurological issue set apart by an unmistakable port-wine stain on the temple, scalp, or around the eye. This stain is a pigmentation brought about by an excess of vessels close to the outside of the skin. Veins on a similar side of the mind as the stain may likewise be influenced.

Sturge-Weber issue is depicted by facial nevus-flames (generally in the scattering of the initial segment of trigeminal nerve anyway not obliged to it), and the recoloring is relied upon to broadened veins in the face that cause the skin to appear to be blushed. In some rare cases, abnormal vessels are not the root cause for any symptoms, but in most cases, it causes symptoms like cognitive impairment, developmental delays, paralysis, feel weakness on one side of their body, etc.

4.2.4   VON HIPPEL-LINDAU (VHL) DISEASE

Von Hippel-Lindau disorder (VHL) is an inherited condition related to tumors emerging in different organs. Tumors in VHL incorporate hemangioblastomas, which are vein tumors of the mind, spinal rope, and eye. They can develop in your cerebrum and spinal string, kidneys, pancreas, adrenal organs, and the conceptive tract.

People with VHL issues conventionally make growths in the kidneys, which leads to tumors, pancreas, and genital tract. They are similarly at an extended risk of working up a kind of kidney threat called clear cell renal cell carcinoma. A pancreatic neuroendocrine tumor is a sort of pancreatic infection that occurs due to VHL disease.

4.2.5   ATAXIA-TELANGIECTASIA (A-T)

Ataxia-telangiectasia (A-T) is an autosomal disorder basically defined using cerebellar degeneration, telangiectasia, immunodeficiency, malignant growth defenselessness, and radiation affectability. A-T is frequently alluded to as a genome precariousness or DNA harm reaction disorder. A-T is an uncommon acquired issue that influences the sensory system, safe framework, and other body frameworks.

It is an autosomal idly obtained illness that has been mapped to chromosome 11q. The extended pace of unpredictable turns of events, vitiligo, odd glucose versatility is viewed. Assessments reveal lessened serum IgA in 3/4th of the patients. Alpha-fetoprotein is commonly used against this disease.

4.3   DEEP LEARNING (DL) CONCEPTS

Convolutional neural networks (CNNs) is the most famous DL architecture, which is applied in analyzing images, audio, and segmented videos. They are enforced in areas where large unorganized data has to be classified. In 2012, the CNN was designed by AlexNet from 8 layers and had exponentially grown to 152 layers of analysis to give the best accuracy. Most of them are applied to the language processor. The goal is to reduce human supervision in feature extraction and classification. CNN is efficient in its computation, which uses convolution and pooling operation to reduce the dimension. This architecture is universally enabled to run on any device for its attraction.

4.3.1   ARCHITECTURE

A simple notation of how a CNN follows a series of layer convolution for a given image is clearly narrated in Figure 4.3.

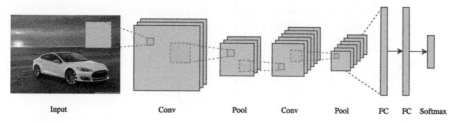

Input   Conv   Pool   Conv   Pool   FC   FC   Softmax

FIGURE 4.3   Architecture of CNN.

When an input image is given into the CNN, the series of operations takes through the convolution, pooling, fully connected layers (FC layers), and the Soft-Max classifier. The layers for the input to the FC are called transition layers, which are normally hidden layers in order to reduce the dimension of the given image.

4.3.1.1   CONVOLUTION

Convolution is the central heart of CNN, which merges two sets of information more precisely. Convolution is an algebraic function to combine the two sets of data, which is achieved on the input image with a filter to produce the output of the convolution layer built as a feature map. It is then fed as the input to the next layer. The first 5×5 matrix is the convolution layer; the 3×3 matrix is the convolution filter, also called the kernel of 3×3 convolution (Figure 4.4).

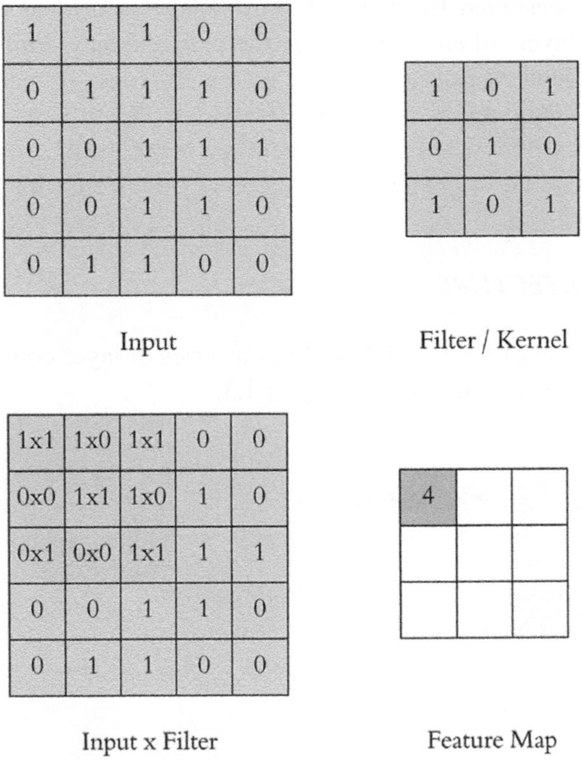

1	1	1	0	0
0	1	1	1	0
0	0	1	1	1
0	0	1	1	0
0	1	1	0	0

1	0	1
0	1	0
1	0	1

Input   Filter / Kernel

1x1	1x0	1x1	0	0
0x0	1x1	1x0	1	0
0x1	0x0	1x1	1	1
0	0	1	1	0
0	1	1	0	0

4		

Input x Filter   Feature Map

FIGURE 4.4   (a & b) Convolution stage-1.

The convolution is achieved automatically by sliding the kernel over the input from first left to right. The sliding is performed continuously from every row-wise matrix based upon the kernel dimension. The entire matrix element has been mapped across the kernel dimensions and summed up. The added sum value goes as the first element of the feature map. The matrix generated by the filter or kernel over the input is known as the receptive-field. Due to the size of the filter, the receptive field is the size of the kernel 3×3. It slides over, and consecutive values are updated in the feature map (Figure 4.5).

1	1x1	1x0	0x1	0
0	1x0	1x1	1x0	0
0	0x1	1x0	1x1	1
0	0	1	1	0
0	1	1	0	0

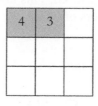

Input x Filter   Feature Map

FIGURE 4.5   Convolution stage-2.

Likewise, the operation performs for the column-wise element, and the feature map is derived. A 2D using a 3×3-filter convolution is shown above. The convolution on input is employed with a different kernel so as to get different feature maps. Later the entire feature map is combined to form an aggregated result. The result of the convolution operation is passed through a hidden layer *ReLU* activation function.

4.3.1.2   RECTIFIED LINEAR UNIT (RELU)

For a neural network, the activation function helps in the transformation of the summed weighted input to the next convolution layer. Whereas the ReLU [7] is a linear function, which will output the input directly when the result is positive, and if there is any other form of output, then it is replaced by zero. This function, in default, helps in neural networks as it is easier to train. Though it is linear, it helps out of complex operations for the data to be learned and avoid the easy situation. This function was not introduced in the early stage of research, and only a recent paper from 2009 started showing head up in detail. The amended linear actuation work is a basic computation that profits the worth gave as information legitimately, or the worth 0.0 if the information is 0.0 or less.

4.3.1.3   BATCH NORMALIZATION (BN)

It is the normalization technique for neural networks to give a standard input to the sub-layer batches. This reduces the batch processing and speeds up the performance of the CNN. As the values are normalized, more processing steps can be reduced. This layer BN with the combination of RLeU is in the transition layers of CNN.

4.3.2 POOLING

The pooling layer is to sink the huge volume of information into an exact worth. Pooling layers decrease the components of the information by joining the yields at one layer into a solitary incentive in the following layer. Nearby pooling consolidates little groups, normally 2×2.

4.3.3 FULLY CONNECTED

After the convolution is combined with pooling layers, we include two or three completely associated layers to wrap up the CNN design. The yield of both convolution and pooling layers are 3D volumes. However, a completely associated layer anticipates a 1D vector of numbers. Along these lines, the yield of the FC must be a solitary vector. As a result, the yield of the last pooling layer is 2×2, the worth must be flatten.

4.4 VISUALIZATION OF CNN

Take a solitary picture x0, which is inputted through a CNN which involves L layers, which in turn executes a change that is non-linear $H_l(.)$, where l specifies to the layer. $H_l(.)$, is an amassed task of layers, for example, batch normalization (BN) [14], rectified direct units (ReLU) [7], Pooling [20], or convolution (Conv). The yield of the lth layer as x_l. Customary convolutional feed-forward systems associate the yield of the lth layer as a contribution to the $(l + 1)^{th}$ layer [17], which offers to ascend to the accompanying layer change.

$$x_l = H_l(x_l-1) \tag{1}$$

Further, the lth layer receives the feature-maps of all preceding layers, $x_0, \ldots\ldots x_{l-1}$, as input:

$$x_l = H_l([x_0, x_1, \ldots . x_{l-1}]) \tag{2}$$

where; $[x_0, x_1, \ldots x_{l-1}]$ alludes to the link of the component maps delivered in layers $0, \ldots l-1$. For quick count, we connect the numerous contributions of $H_l(.)$, in Eqn. (2) into a solitary tensor (Figure 4.6).

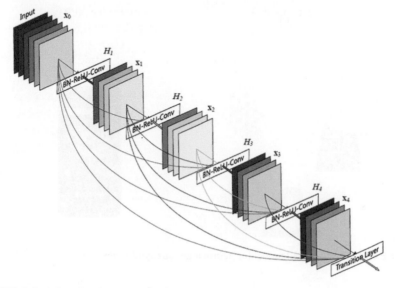

FIGURE 4.6   Visualization of CNN layers.

Composite capacity-propelled by Ref. [14], we characterize H(.) as a composite capacity of three back to back tasks: cluster standardization (BN) [15], trailed by a corrected straight unit (ReLU) [7] and a 3×3 convolution (Conv). Pooling layers-Eqn. (2) is a link activity of the considerable number of layers to down example and change the size of highlight maps. The progress layers utilized in our trials comprise of a BN layer and a 1×1 convolutional layer followed by a 2×2 normal pooling layer.

4.5   CCNN-BASED MODEL

The new technique called cloud-based convolution neural network (CCNN) has been proposed. This work is purely an online service provider for the test results to be updated earlier. This has been evolved from the CAD system proposed by many researchers so far. Normally the neurocutaneous syndromes are not only analyzed by the MRI images but also by the clinical findings (Figure 4.7).

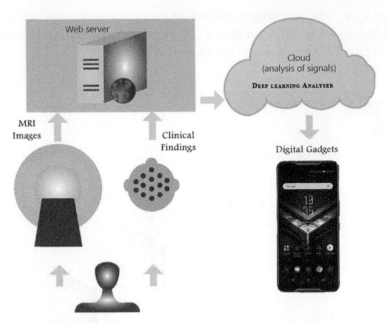

FIGURE 4.7   CCNN based model for neurocutaneous syndromes.

Thus, as the case is by both image and text type of data, a special architecture to handle both types of data is proposed. From Figure 4.7, both the MRI images and clinical findings are feed into the webserver, which is the repository of the patient's data. The data from the webserver is accessed by the special cloud deep learning analyzer (DLA) for the result of the patient report. The major role of our architecture is all the DL analysis is made inside the cloud. The mobile phone, which is at high usage, retrieves all information about the patient from the cloud-based upon some unique identifier.

MRI, which is known for magnetic resonance imaging, is an image born from the soft tissue of the human body. The images are created by using a magnetic field with the help of radio waves over it. As it is obtained by using the axis points, the produced image has detailed pixels where it can show the smallest abnormality. There are certain symptoms that can be found out by the clinical findings by the physicians, which further intend to MRI for the confirmation of the diseases. There are sometimes, by clinical findings itself certain diseases can be sorted out. These clinical findings can be reported by the doctor. Adding both the data as input to the server will actually give the best results for any type of disease. Thus, as a result, the output achieved by this model is 99.9% accurate. The DLA takes responsibility in formulating the complete idea. The normal convolution theory is followed by the last decision factor called the decision tree algorithm is added to confirm the syndrome registered by the physician's report. Using the MRI and the clinical finding, the syndrome can be easily confirmed.

4.5.1 DEEP LEARNING ANALYZER (DLA)

Preprocessing is a technique followed in the input data before entering into the processing steps. The redundancies are removed and then feed into the CNN (Figure 4.8).

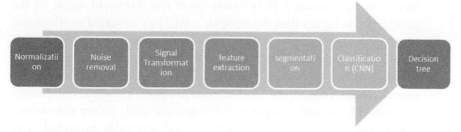

FIGURE 4.8   DLA methodology.

- **Step 1: Normalization:** Any form of data is good to be normalized as all the pixel of the image is not suited. The noise-like illumination and other factors in the MRI scan images while capturing is filtered first. SEB, commonly known as selected and estimated method motivated by Ref. [24], is used for normalization. This normalization technique finds the best line of pixel to compute the least square, which is based on the SEB of the nearby pixels.
- **Step 2: Noise Removal:** Weiner filter is the common filter that is used to reduce the noise. It eliminates the noise from the normalized image of Step 1. It protects the inner details and smooths the crisp edges of the image. The anisotropic filter can also be tried motivated by Ref. [25]. Finally, based upon the PSNR values, the former provides lower PSNR than the later. The Weiner filter is commonly used for noise removal.
- **Step 3: Signal Transformation:** The image transformation is done to remove the repeated information, and then the features are extracted. Discrete wavelet transforms (DWT) is a majorly used technique for

numerical analysis and functional analysis, in any wavelet transform for which the wavelets are discretely sampled. It is used to convert the signal from the low-frequency domain to the high-frequency domain. As a result, it helps in exact feature extraction.

- **Step 4: Feature Extraction:** Curvelets are a non-adaptive technique for multi-scaling object representation. Thus curvelet transforms with higher DWT are helpful in representing images from different angles too. The curvelet transform is one of the computer vision and imaging technique where a higher dimensional DWT represents the images in multiple angles and scales.

- **Step 5: Segmentation:** The gray to binary image conversion is based on thresholding. This is the simplest form of segmentation, especially based on pixel values. It is based upon the threshold value of the pixels. If it is higher than the normal, it will be converted as white and vice versa. The image with redundant information drops the performance of the computational result. Therefore, feature extraction helps in the removal of repeated patterns, which will not impact performance much. Thus, linear discriminant analysis, principal component analysis (PCA), and independent component analysis are the various dimensionality reduction techniques used generally preferred. The commonly enhanced version of PCA is employed famously known as kernel PCA.

- **Step 6: Classification (CNN):** The two important factors in classification are training and testing. Whatever data is fed into classification has to be trained well with the already available data sets. The CNN working maps into this section. They follow a series of layers from convolution, pooling, and connected layers. The commonly used classifiers are SoftMax, SVM, Naïve-Bayes–Gaussian mixer model, random test, KNN, and decision tree.

- **Step 7: Decision Tree Algorithm:** This has emerged from the family of supervised learning algorithms [3]. This algorithm is most suitable for decision making when there are many symptoms to predict a syndrome (Figure 4.9).

Based upon the root node, the classification begins, which is mapped with the symptoms of the clinical findings. The node further refines the decision-making of whether the syndrome is found or not with the terminal node.

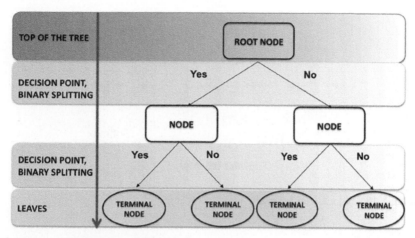

FIGURE 4.9   Decision tree algorithm.

4.5.2 DISCUSSION

Convolutional neural network (CNN, or ConvNet) is a special kind of multi-layer computer vision neural network, which is designed for recognizing patterns from the image pixels directly with a very low preprocessing steps, it has been one of the popular neural analyzers. There are many built-in online CNN architecture to analyze, such as LeNet, AlexNet, VGG, GoogLeNet, ResNet. GooglLeNet has achieved a top-5 error rate of 6.67%. Thus, our proposed architecture has been deployed with GooglLeNet. It has a 1×1 convolution for dimension reduction, which in turn increases the performance. This reduces bottleneck computation. Eventually, depth and width can be increased. Thus implements CNN in a more accurate fashion (Figure 4.10).

Based upon the learning from the GoogLeNet the Learning is transferred to the proposed block, which is the DLA in the cloud.

Figure 4.11 shows the input of the training data, which is taken from the Figshare repository for testing and then fed into the classifier. The output of this framework gives the complete list of the symptoms and findings in the predicted table. Based on the table input, the decision tree algorithm works and finds the excellent decision of whether the syndrome occurs or not. The visual preprocessing stage of the MRI image is demonstrated in Figure 4.12. The CNN layer visualization is shown in layers.

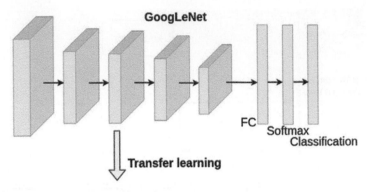

FIGURE 4.10   GoogLeNet schematic view.

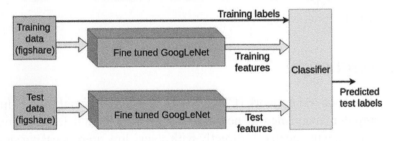

FIGURE 4.11   Data preprocessing in GoogLeNet.

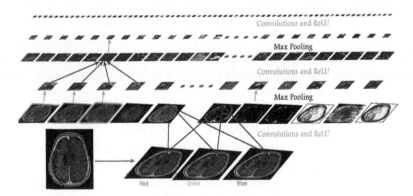

FIGURE 4.12   General overall framework for syndrome classification.

The proposed CCNN model is implemented in MATLAB-2018b, and compared to the traditional classification system, this approach has accounted for almost 99% accuracy.

4.5.3 DISCUSSION

Once when the patient's MRI and the clinical findings are fed into the cloud. The DL analyzer in the cloud performs the layer performance based upon the image size. The CNN architecture works as per the GoogleLeNet operation to complete the image analysis and feeds the data to the decision tree classifier. This classifier has the conditional probability symptoms to the disease for the correct identification of the syndrome. All these operations will be performed in a fraction of a second. This type of CCNN architecture will be more useful when large data set MRI is under pending without the diagnosis. This architecture is a general prototype for all types of medical imaging. This model is the only novelty for the MRI images as of now. Future work can be extended based upon the requirement.

4.6 CONCLUSION

Thus, a new innovative approach of cloud-based DL has been proposed in this chapter. The more detailed version can be imposed in further sections. The original concept of CCNN has been proposed for the live sharing of information to the globe. This system is proposed specific to one particular syndrome called neurocutaneous syndromes. Further, this approach can be interlinked with the online patient monitoring system in the future to support full medical care.

KEYWORDS

- **artificial intelligence**
- **batch normalization**
- **cloud computing**
- **cloud-based convolution neural network**
- **convolutional neural network**
- **neurological disorders**

REFERENCES

1. Mamta, M., Lalit, M. G., Sumit, K., Iqbaldeep, K., Amit, V., & Jude, H. D., (2019). Deep learning based enhanced tumor segmentation approach for MR brain images.

Applied Soft Computing, 78, 346–354. ISSN: 1568-4946, https://Doi.Org/10.1016/J. Asoc.2019.02.036.

2. Deepak, S., & Ameer, P. M., (2019). Brain tumor classification using deep CNN features via transfer learning. *Computers in Biology and Medicine, 111,* 103345. ISSN: 0010-4825.

3. Arun, M. S. S., & Rajesh, R. S., (2012). An effective spam filtering for dynamic mail management system, *ICTACT Journal on Soft Computing, 2*(3), ISSN: 2229-6956(online). doi: 10.21917/ijsc.2012.0050.

4. Lee, D., Yoo, J., Tak, S., & Ye, J. C., (2018). Deep residual learning for accelerated MRI using magnitude and phase networks. In: *IEEE Transactions on Biomedical Engineering,* (Vol. 65, No. 9, pp. 1985–1995). doi: 10.1109/TBME.2018.2821699.

5. Raghavendra, Acharya, U. R., & Adeli, H., (2019). Artificial intelligence techniques for automated diagnosis of neurological disorders. *Eur. Neurol., 82,* 41–64. doi: 10.1159/000504292.

6. Huang, G., Liu, Z., Van, D. M. L., & Weinberger, K. Q., (2017). Densely connected convolutional networks. In: *2017 IEEE Conference on Computer Vision and Pattern Recognition (CVPR)* (pp. 2261–2269). Honolulu, HI. doi: 10.1109/CVPR.2017.243.

7. Glorot, X., Bordes, A., & Bengio, Y., (2011). *Deep Sparse Rectifier Neural Networks.* In AISTATS.

8. Goodfellow, I., Warde-Farley, D., Mirza, M., Courville, A., & Bengio, Y., (2013). *Maxout Networks.* In ICML.

9. Gross, S., & Wilber, M., (2016). *Training and Investigating Residual Nets.* http://torch. ch/blog/2016/02/04/resnets.html.

10. Hariharan, B., Arbeláez, P., Girshick, R., & Malik, J., (2015). *Hyper Columns for Object Segmentation and Fine-Grained Localization.* In CVPR.

11. He, K., Zhang, X., Ren, S., & Sun, J., (2015). *Delving Deep into Rectifiers: Surpassing Human-Level Performance on Image Net Classification.* In ICCV.

12. He, K., Zhang, X., Ren, S., & Sun, J., (2016). *Deep Residual Learning for Image Recognition.* In CVPR.

13. He, K., Zhang, X., Ren, S., & Sun, J., (2016). *Identity Mappings in Deep Residual Networks.* In ECCV.

14. Huang, G., Sun, Y., Liu, Z., Sedra, D., & Weinberger, K. Q., (2016). *Deep Networks with Stochastic Depth.* In ECCV.

15. Ioffe, S., & Szegedy, C., (2015). *Batch Normalization: Accelerating Deep Network Training by Reducing Internal Covariate Shift.* In ICML.

16. Krizhevsky, A., & Hinton, G., (2009). *Learning Multiple Layers of Features from Tiny Images.* Tech Report.

17. Krizhevsky, A., Sutskever, I., & Hinton, G. E., (2012). *ImageNet Classification with Deep Convolutional Neural Networks.* In NIPS.

18. Kumar, S., Dabas, C., & Godara, S., (2017). Classification of brain MRI tumor images: A hybrid approach. *Procedia Comput. Sci., 122,* 510–517.

19. Deepak, S., & Ameer, P. M., (2019). *Computers in Biology and Medicine, 111, 103345.*

20. Mohan, G., & Subashini, M. M., (2018). MRI based medical image analysis: Survey on brain tumor grade classification. *Biomed. Signal Proces., 39,* 139–161.

21. Yousefi, M., Krzyzàk, A., & Suen, C. Y., (2018). Mass detection in digital breast tomosynthesis data using convolutional neural networks and multiple instance learning. *Comput. Biol. Med., 96,* 283–293.

22. Gu, Y., Lu, X., Yang, L., Zhang, B., Yu, D., Zhao, Y., & Zhou, T., (2018). Automatic lung nodule detection using a 3D deep convolutional neural network combined with a multiscale prediction strategy in chest CTs. *Comput. Biol. Med., 103*, 220–231.

23. Zuo, H., Fan, H., Blasch, E., & Ling, H., (2017). Combining convolutional and recurrent neural networks for human skin detection. *IEEE Signal Process. Lett., 24*(3), 289–293.

24. Charron, O., Lallement, A., Jarnet, D., Noblet, V., Clavier, J. B., & Meyer, P., (2018). Automatic detection and segmentation of brain metastases on multimodal MR images with a deep convolutional neural network. *Comput. Biol. Med., 95*, 43–54.

25. Shao, L., Zhu, F., & Li, X., (2015). Transfer learning for visual categorization: A survey. *IEEE Trans. Neural Netw. Learn. Syst., 26*(5), 1019–1034.

26. Zhou, L., Zhang, Z., Chen, Y. C., Zhao, Z. Y., Yin, X. D., & Jiang, H. B., (2019). A deep learning-based radiomics model for differentiating benign and malignant renal tumors. *Transl. Oncol., 12*(2), 292–300.

27. Deniz, E., Şengür, A., Kadiroğlu, Z., Guo, Y., Bajaj, V., & Budak, Ü., (2018). Transfer learning based histopathologic image classification for breast cancer detection. *Health Inf. Sci. Syst., 6*(1), 18.

28. Hussein, S., Kandel, P., Bolan, C. W., Wallace, M. B., & Bagci, U., (2019). Lung and pancreatic tumor characterization in the deep learning era: Novel supervised and unsupervised learning approaches. *IEEE Trans. Med. Imaging.* https://doi.org/10.1109/TMI.2019.2894349.

29. Liu, R., Hall, L. O., Goldgof, D. B., Zhou, M., Gatenby, R. A., & Ahmed, K. B., (2016). Exploring deep features from brain tumor magnetic resonance images via transfer learning. *IEEE International Joint Conference on Neural Networks (IJCNN)*, 235–242.

30. Ahmed, K. B., Hall, L. O., Goldgof, D. B., Liu, R., & Gatenby, R. A., (2017). Fine-tuning convolutional deep features for MRI based brain tumor classification. *International Society for Optics and Photonics Medical Imaging 2017: Computer-Aided Diagnosis, 10134*, 101342E.

31. Yang, Y., Yan, L. F., Zhang, X., Han, Y., Nan, H. Y., Hu, Y. C., & Ge, X. W., (2018). Glioma grading on conventional MR images: A deep learning study with transfer learning. *Neurosci., 12*.

32. Talo, M., Baloglu, U. B., Acharya, U. R., (2019). Application of deep transfer learning for automated brain abnormality classification using MR images. *Cogn. Syst. Res., 54*, 176–188.

33. Jain, R., Jain, N., Aggarwal, A., & Hemanth, D. J., (2019). Convolutional neural network-based Alzheimer's disease classification from magnetic resonance brain images. *Cogn. Syst. Res.* https://doi.org/10.1016/j.cogsys.2018.12.015.

34. Swati, Z. N. K., Zhao, Q., Kabir, M., Ali, F., Zakir, A., Ahmad, S., & Lu, J., (2019). Content-based brain tumor retrieval for MR images using transfer learning. *IEEE Access, 7*, 17809–17822.

35. Cheng, J., Huang, W., Cao, S., Yang, R., Yang, W., Yun, Z., & Feng, Q., (2015). Enhanced performance of brain tumor classification via tumor region augmentation and partition. *PLoS One, 10*(10), e0140381.

36. Cheng, J., Yang, W., Huang, M., Huang, W., Jiang, J., Zhou, Y., & Chen, W., (2016). Retrieval of brain tumors by adaptive spatial pooling and Fisher vector representation. *PLoS One, 11*(6), e0157112.

37. Ismael, M. R., & Abdel-Qader, I., (2018). Brain tumor classification via statistical features and back-propagation neural network. *IEEE International Conference on Electro/ Information Technology, (EIT)*, 0252–0257.

38. Abiwinanda, N., Hanif, M., Hesaputra, S. T., Handayani, A., & Mengko, T. R., (2018). Brain tumor classification using convolutional neural network. Springer *World Congress on Medical Physics and Biomedical Engineering*, 183–189.

39. Pashaei, A., Sajedi, H., & Jazayeri, N., (2018). Brain tumor classification via convolutional neural network and extreme learning machines. *IEEE 8th International Conference on Computer and Knowledge Engineering, ICCKE*, 314–319.

40. Afshar, P., Plataniotis, K. N., & Mohammadi, A., (2019). Capsule networks for brain tumor classification based on MRI images and course tumor boundaries. *IEEE International Conference on Acoustics, Speech and Signal Processing, ICASSP*, 1368–1372.

41. Deepak, & Ameer, P. M., (2019). *Computers in Biology and Medicine, 111*, 103345.

42. http://www.Towardsdatascience.com (accessed on 18 December 2020).

43. https://radiopaedia.org/ (accessed on 18 December 2020).

CHAPTER 5

CRITICAL INVESTIGATION AND PROTOTYPE STUDY ON DEEP BRAIN STIMULATIONS: AN APPLICATION OF BIOMEDICAL ENGINEERING IN HEALTHCARE

V. MILNER PAUL,[1] S. R. BOSELIN PRABHU,[2] T. JARIN,[3] and T. ANANTH KUMAR[4]

[1]Research Scholar, Department of Electrical Engineering, National Institute of Technology, Manipur, India, E-mail: vithayathilmilner@nitmanipur.ac.in

[2]Department of Electronics and Communication Engineering, Surya Engineering College, Mettukadai, Tamil Nadu, India, E-mail: eben4uever@gmail.com

[3]Department of Electrical and Electronics Engineering, Jyothi Engineering College, Thrissur, Kerala, India, E-mail: jeroever2000@gmail.com

[4]Department of Computer Science and Engineering, IFET College of Engineering, Tamil Nadu, India, E-mail: ananth.eec@gmail.com

ABSTRACT

Electrical stimulation of the brain is a vital tool for the examination of brain activities on a neural basis. It is a fruitful method for exploring the brain-behavior relationships. Studies show that there is a considerable response when triggering pulses are applied in specified regions. Neurons that can be triggered successively can be connected. Multineuron stimulation is a unique method that stimulates the brain in different areas. For this, introduce a product called "Brain Stitcher" that helps to broaden the capability of the brain. Brain stitcher can make the connection between the desired parts of the brain. Depending upon the neurologists' recommendation, the proper amount of voltage for the proper amount of time could be applied to a series of locations in the brain repetitively by this brain stitcher, which in turn will wire the required parts of the brain. This will be a promising product that can be considered a cure for various diseases that are closely about brain activity. It includes conditions like mental retardation, epilepsy, Parkinson's disease, cognitive disabilities, neural disorders, brain stroke, and even for normal people.

5.1   INTRODUCTION

In this fast-developing and collectively well-versed present-day entire world, people, in general, keep remarkable desires concerning one's entire potential that features both the mental and physical workouts in our system. Assessing today's speed related to advancing the individuals along with intense as well as mediocre cerebral possibilities are undoubtedly deemed highly effective nationals which actually in a natural way discharges about differently-abled as well as demolish each of them become damaging as well as culturally regarded as throwing away human resources. It can be projected that 7–8% of the net whole world inhabitants include the disabled people additionally in just 75–90% are intellectually questioned. In addition, the rate of terminating the pregnancy is, in fact, a great number of because of the confirmation of those cerebral disabilities. The worldwide recent statistics indicate overall frightening tendencies. However, this is detrimental; a faulty brain can assist a person to learn the typical cerebrum. Upon the basic facts obtained thus far, numerous tries have already been provided enhance the methods for mentally difficulty young children [1].

This product strives over at designing a highly innovative piece of equipment materialized just like a biochip which should stimulate the health of the brain via a digital signal and thus assist in enhancing the potential of one's brain. All of this product can be used to stitch the brain. The comprehensive procedure can easily be known as stitching over time. Merely by influencing happenings, time relationship can be produced within the brain, clearly as the brain can line collectively the neurons which happen to be firing in instant succession [2]. To make these types of interconnection, a chain generator or perhaps even *stitcher* over time may be used. The necessary excitements must be applied to the expected areas of the brain about necessary momentum. Thus parts of the brain can easily be introduced closer over time. In addition, this is known as a good 'stitch in time' considering the brain. Typically, doctors stitch in space carrying a couple of pieces nearer in space, but here it can be stitch overtime delivering a couple of parts nearer over time [3].

5.2 DEFINING THE CONCEPTS AND LITERATURE SURVEYS

5.2.1 ANATOMY OF BRAIN

The brain is undoubtedly an astounding three-pound system that in fact manages every processes of one's body system, determines the essential features of important information beginning with the outside world as well as embraces overall the aspect your mental state and certainly life force (Figure 5.1).

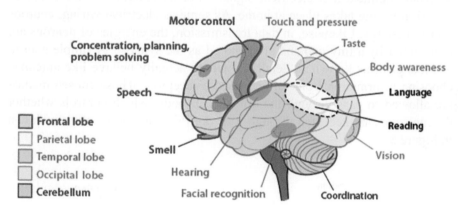

FIGURE 5.1   Anatomy of brain.

Five senses which are used to receive information are touch, sight, taste, smell, and by listening. It usually clubs the meaning in such a way that it has an appropriate meaning that we can comprehend and at the same time, it can store data and as well as information in our faculty of memory. Our thinking pattern, speech, memory, and other functions are controlled by the brain. It also determines how the responses are in traumatic circumstances (such as taking an exam, dropping a trade, or misery) by controlling the heart and breathing rate [8].

5.2.1.1 NERVOUS SYSTEM

The nervous systems are classified into peripheral and central systems. The central nervous systems are denoted as CNS and are composed of the spinal cord and brain. The peripheral nervous system is denoted as PNS and is composed of spinal nerves that branch from the spinal cord and cranial nerves that branch from the brain. As a whole, brain is a more complicated structure, as a fact, scientists are still doing their research to confirm almost for the new invention. Comparing to other the range of the brain's potentials is not known but in fact it is in one of the complicated system in the entire systems [5].

5.2.1.2 NERVE CELLS

The neurons are of different shapes and sizes, consisting of an axon, cell body, and dendrites. In neurons, the information transformations take place utilizing chemical and electrical signals. Let us imagined the concept of electrical wiring takes place at home. As same as electrical wiring, neurons transmit energy. Likewise, in data transmission, the energies of neurons are transmitted by a tiny gap and are denoted as the synapse. A simple neuron has lots of arms known as dendrites, which actually behave like antennae choosing information from different nerve receptors. These chosen memos are allowed to pass through the minute cell body, which controls whether the messages are accepted through it or not [6]. The nerve cells are shown in Figure 5.2.

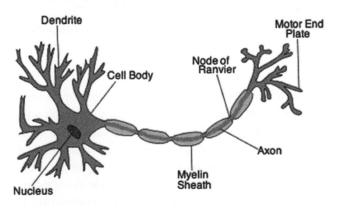

FIGURE 5.2   Nerve cells.

5.2.1.3   NEUROPLASTICITY

The nerve cells are also known as neurons. Neuroplasticity is used to compensate for injury or disease-affected brain to adjust its activities according to the situations. The mechanisms used for reorganization is referred to as "axonal sprouting." The undamaged axons are used for growing new nerve and to reconnect the neurons. The sprout nerve endings of the undamaged axons are connected with undamaged nerve cells. Thus it forms a new neural pathway for needed functions.

After the advanced scientific study of the human brain system has paved the way for leading to the evolution of various artificial intelligence (AI) techniques like neural network system like multilayer perceptron (MLP); now the day is neural systems are the most operative and fitting AI technology. The vital concerns in multilayer perceptrons (MLP) design comprise a description of the amount of unseen deposits and the quantity of units in these stratums. Here, multilayer perceptrons (MLPs) remain covered feed-forward systems classically qualified with stationary backpropagation. Diverse methods for veiled neurons assortment such as simple method, based on hopfiled neural network, Akaike's information criterion (AIC), inverse test error method, depend on neural network error, depend on neural network training, and depend on neural network output, hidden layers, and MSE [5].

5.2.2   INTRACRANIAL SELF-STIMULATION

Research exploring about gratifying characteristics related to intracranial self-activation (ICSS) has noted dopamine as the neurotransmitter concerned about incentive. Zones that helps ICSS recognized like pleasure focus, overrun by using recognized dopamine techniques about the cerebrum. The dopamine structures concerned in incentive is overall the mesotelencephalic dopamine structure (referred in to like overall the midbrain dopamine structure). It is based on dopaminergic neurons scheduling of the mesencephalon (the midbrain) in several zones concerning the telencephalon. The neurons that often put together all of this structure include their cell system in the substantia Nigra as well as the ventral tegmental areas [1].

About managing much related to the input part of these neuronal networks. ICSS offers a simple exclusive aid in neuropharmacological exploration to examine the effect of different components relating to incentive as well as reinforcement procedures. Intracranial self-stimulation varies substantially from drug self-managing concerning that, in this processes, the animal is functioning directly stimulate assume reinforcement circuitry in the cerebrum and the consequences related to drugs are evaluated relating to these incentive thresholds. Drug use of exploitation reduces the tolerance for ICSS, and generally, there is a top-notch interaction within the capability of drugs to reduce ICSS tolerances as well as their exploitation [1].

5.2.2.1   ICSS: A MODEL TO STUDY NEURONAL PLASTICITY

Only recently, the electrical self-stimulation archetype already has turned out to be incredibly beneficial in delineating about neural substance concerned concerning researching as well as medication exploitation. Self-stimulation includes operant research, which actually can easily stimulate alterations in the neuronal cytoarchitecture. Appropriately we include utilized the experimental archetype to research study learning associated neuronal plasticity. We include carried out a combination of investigations to effectively establish self-stimulation (SS) gratifying experience activated neural plasticity in overall the hippocampus as well as motor cortical neurons [7].

Overall the electrical self-stimulation already has been deemed as one of overall the powerfully pleasing conduct encounter, quite possibly even much more prominent compared to nourishing or sexual incentives. Concerning add-on to dendric development, a considerable enhance is viewed concerning the density of lacunosam, radiatum, and lucid laminae concerning the CA3 zone of the hippocampus.

ICSS triggered synaptogenesis during the CA3 zone, which comprises sub-atomic, radiatum, and lucidium components of the hippocampus and sub-atomic level of the hippocampus and motor cortex. Also, SS outcomes in raise about the stages of glutamate, noradrenaline, dopamine, as well as improvement of activity in the hippocampus and the motor cortex. Additionally, the before investigations are acknowledged to enable the acquisition of operant as well as spatial researching activities about rats. This sort of facilitation could be due to a rise in dendritic amortization linked along with neurochemical compound alterations in the hippocampus [8].

5.2.2.2   ICSS: AN ANIMAL MODEL FOR DEPRESSION

ICSS archetype offers a functional assessment of anhedonia, a main function of depression. Since of anhedonia encountered by depressed individuals highly recommend that each of these persons could reveal alternations about incentive procedure, ICSS procedure already has remained suggested as a kind of depression treatment. Significant facts imply that ICSS thresholds are trustworthy calculate of incentive that displays the whole continuum from hedonic to anhedonia.

The research study of neural substrates related to ICSS conducts subsequent prototype or medicinal manipulations guarantees to stimulate our knowledge of gratifying processes that seem to be to become modified about several psychological issues, such as depression [9].

Two manipulations remained to give out an anhedonic status in creatures, as an operationally outlined by using regulates in ICSS reactions rate related to rates of thresholds: (a) exposure to effectively uncontrolled tension as well as; (b) withdrawing of prolonged-term experience to psychomotor inciters. Tricycles show up to become useful concerning getting rid of the consequences related to withdrawing of amphetamine relating to ICSS. However, several more research studies usually are required, to facts to date implies that the ICSS model has got the wonderful form and certainly etiological validity and displays medicinal similarities as a version for the medication overused anhedonia. However, the specific involvement within drug elicited as well as on medication-activated depression concerning humans is not recognized. For that reason, the practical etiological authenticity related to overall the ICSS archetype just as a version of the non-medication stimulated desolation continues to be vague. Forthcoming scientific and preclinical investigation wishes to tackle all of this concern furthermore [10].

5.2.2.3   GENERAL METHODOLOGY

The notion of about work approach is mature male Wistar lab rat's body-weight the variability 250–275 grams are employed to use on the stereotaxic implantation that is bipolar electrodes within the substantia Nigra-ventral tegmental areas (SN-VTA) bilaterally. Overall the equipment put to use is Stereotaxic table, surgical procedure equipment, the stimulator (programmable pulse generator), Skinner's box transformed to have ICSS as well as networking wires and cables. About surgical treatment explains overall, the rat is anesthetized along with sodium pentobarbitone (40 mg/kg) and after that, set in the stereotaxic body along with rat adaptor. After a shot of lignocaine anesthesia directly into the scalp zone, the skull surface appeared to be open to obtaining sites. Flat-skull coordinates applied extracted from

Paxinos, and Watson rat drawing usually is signified by using bregma as a reference aspect as well as burr openings are drilled throughout the skull. The bipolar electrodes usually are fixed continually right into SN-VTA [1].

The stereotaxic coordinates for SN-VTA are as follows:

- **Antero-Posterior (AP):** –3.5 to 4.5 mm;
- **Medio-Lateral (ML):** 1.1 to 1.8 mm;
- **Dorso-Ventral (DV):** 8.5 + 0.2 mm.

ICSS behavioral response is used for testing the surgical rat after 5–7 days. The electrodes placed in the head are connected to the stimulator and the rat is placed in the testing chamber. The testing chamber is also called the Skinner's box, the stimulator as a pulse generator. The Skinner's box has a pedal (lever) on one of the walls of the chamber that is connected to the microswitch, which has a connection with the pulse generator. While pressing the lever, it delivers pulse current by completing the circuit process. Figure 5.3 shows the intracranial self-stimulation structure.

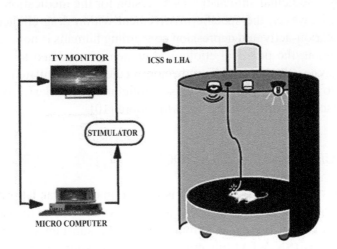

FIGURE 5.3   Intracranial self-stimulation.

As the present current range is noticed which often works as an incentive, influencing may be used in to rapidly set up switch pressing; i.e., merely by reinforcing primarily motions, after that locomotion in the direction of the switch, after that sniffing all around the switch, after that coming in contact with and eventually pressing it. Quickly the rat gains knowledge to forecast this relationship related to pedal pressing and of course the gratifying activation. By using rat's response, frequency, and current strength are monitored for stimulus [7].

5.2.3 HEBBIAN THEORY

Hebbian theory is formulated by Donald Hebb in 1949. The Hebbian theory comes under neuroscience. The structure of the Hebbian theory is shown in Figure 5.4. The Hebbian theory states that the neuron adaptations are organized based on its behavior. It is also called Hebb's rule or cell assembly theory [5].

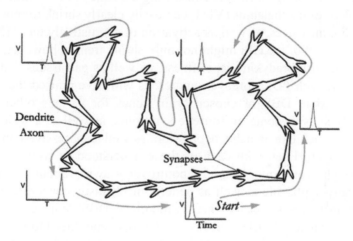

FIGURE 5.4   Hebbian theory.

5.2.4 DEEP BRAIN STIMULATION (DBS)

Deep brain stimulation (DBS) already has offered outstanding advantages with regards to people in general, along with a diversity of neurologic circumstances. Activation related to the ventral moderate nucleus of the thalamus can easily drastically lessen tremors linked with vital tremors or sometimes Parkinson's disorder (PD). Also, activation of those the subthalamic nucleus (STN) or sometimes the internal segment of the globus pallidus (GPi) can easily elegantly scale down bradykinesia, stiffness, tremor, as well as gait issues concerning people along with PD. Numerous groups are striving to prolong this type of treatment method to various other issues. Nonetheless, the specific technique of task of DBS continues to be undecided. These types of research studies include significance that stretches more than medical therapeutics. Explorations of the techniques related to the task of DBS may have the possibility to shed light on quintessential concerns these types as the functioning physiology of designated brain connections as well as the involvement within endeavor concerning all those connections as well as behavior. However, we take a look at applicable medical concerns; all of us highlight the significance of present and forthcoming studies relating to these matters [11].

DBS is known for having offered great medical gain for people in general along with crucial tremor (ET) and PD. The positioning of that intense frequency stimulating electrodes concerning the zone of the ventral interme-diate nucleus of the thalamus (VIM) can easily clearly shrink tremor in each of these circumstances, as well as activation of alternatively the STN or the internal segment of the GPi might not only shrink tremor, however, besides but also diminish bradykinesia, stiffness, as well as gait issues that plague people along with PD. Moreover, numerous will have praised the possible reward-related to DBS of chosen brain zones for various other change illnesses such as dystonia or Tourette syndrome, as well as a wide variety of problems such like an ache, depression, as well as obsessive-compulsive disorder (OCD). Even with each of these understood as well as possible breakthroughs concerning therapy, controversy swirls about a variety of scientifically applicable as well as general mechanistic concerns. Precisely what circumstances are open to therapy merely by DBS? What are overall the processes of an act of DBS? Precisely what outcome does DBS include on operate of the brain connections? We tackle each of these disputed concerns and stress require for the forthcoming studies. To put the step, nonetheless, we first take a look at the background of overall the progress of DBS as a beneficial resource [12].

5.2.4.1   HISTORY OF DEEP BRAIN STIMULATION (DBS)

Always since the conventional presentation of the local electrical excitability of the motor cortex, electrical activation of the cerebrum has performed a huge part in studies of cerebrum function. They primarily a document of human cortical activation made an appearance 4 years down the road. While electrical activation appeared to be utilized to map cortical operation in the 1930s, it appeared to be not really until human stereotaxic instruments have been created that neurosurgeons could start to examine the results of revitalizing closer configurations. Merely by at the beginning of the 1950s, intraoperative activation turned out to be utilized to determine profound configurations such as the corticospinal system before to lesioning the GPi or thalamus. Many results in the 1950s concentrated on optimistic phenomena that have been taken from merely by activation. In the early 1960s, it was confirmed that high frequency (100-Hz) activation of the ventrolateral thalamus may reduce tremor.

The notion of tackling neurologic diseases with persistent stimulation commenced to arise in the 1960s; however, activation was mostly utilized for focusing on surgical procedure lesions. Formed a technique of implanting a bunch of numerous electrode electrical wires immersed in overall the cerebrum as well as going out of all of them in position for several weeks, throughout which activation could get served. The goal and purpose of the activation appeared to be to demarcate the "best" goal for a successive lesion. Along with the fixed wires, a lesion might be created in minimal actions more than the duration of days to weeks to attempt to accomplish the highest possible benefit without ever-unseemly results. Despite the fact overall the goal and purpose appeared to be nonetheless lesion assistance, this is quite possibly overall the first document of activation via recurrently fixed electrodes.

During the early 1970s, results of utilizing persistent activation thera-peutically evolved with regards to remedying ache, gesture problems, or epilepsy. Issued the primarily huge group of persistent cerebellar activation research studies with regards to cerebral palsy. During all those instances, activation was delivered transcutaneously via inductive coupling instruments to electrodes fixed on the external related to the cerebellar cortex. Improve-ment was reported to take place in 49 of 50 people. Nevertheless, cerebellar activation in cerebral palsy subsequently lessened out of preferring if blinded research studies did not exhibit regular advantages. By 1980, various other research results of handling movement diseases along with persistent activa-tion had made an appearance. However, the primarily lasting internally fixed cardiovascular pacemaker appeared to be formed by 1960, it appeared to be not up to the 1990s that implantable pacemaker solutions were merged with recurrently fixed deep brain electrodes with regards to lasting persistent DBS. Considering that then, DBS has come to be extremely utilized for controlling a diversity of diseases.

5.2.4.2   DEEP BRAIN STIMULATION (DBS) FOR ESSENTIAL TREMOR

The initial significant use of DBS within the United States and the EU (Europe) was in fact for the remedy for ET or the tremor-related to PD. Primarily confirmed the usefulness of VIM activation along with implantable pulse generators. Then, they confirmed a more substantial sequence of patients along with VIM activation regarding the treatment for tremor, along with considerable gain within the majority of patients. Future single, as well as multicenter research studies, have regularly confirmed the significant advantage of VIM activation with regards to ET using an ordinary tremor reducing well over 80% within the considerable number of patients [1].

5.2.4.3   DEEP BRAIN STIMULATION (DBS) WITH REGARDS TO PARKINSON DISORDER (PD)

Diverse services of activation provide you with varying medical outcomes in PD. Thalamic stimulation inside the area of overall the VIM activation cut down limb tremor but small effect located on various other manifestations considering the disease. Stimulation of the GPi might cut down every one of the huge motor manifestations of PD, which includes alleviation of dopa-induced dyskinesias, unconscious motions provided by specific dosages of dopaminergic medicinal drugs that could restrict therapy usefulness. GPI activation also might cut down painful spasms as well as sensory signs and symptoms that could happen whenever the gain from specific dosages related to levodopa abates. Nonetheless, GPi stimulation doesn't generally sanction overall the alleviation of drugs, and such could be a severe restriction for those who run drug-induced negative effects, for instance, orthostasis, psychosis, daytime weariness, or cognitive problems. STN DBS offers comparable alleviation of motor signs and symptoms. Some research studies show that mutual STN DBS greatly enhances gait, tremor, and bradykinesia and also licenses overall the alleviation of dopaminergic medicines, causing less drug-induced side effects. Direct, unrestrained side by side comparisons related to GPi DBS along with STN DBS has already been carried out, except a preliminary document associated with a managed evaluation of the gain from GPi DBS in contrast with STN DBS ensures overall the compared medical gain from activation over at either site along with small modify concerning preoperative medicines in the GPi association instead of the STN group [2].

The level of gain from STN DBS or GPi DBS doesn't typically go over that discovered from individual dosages related to levodopa in every affected person. However, DBS offers two primary benefits: (a) it minimizes the amount of time a patient uses within the "off" area whenever the gain from an individual amount of drugs already has diminished—for some, this off area causes someone gradual, unstable, rigid, as well as not able to climb coming from a chair, as well as; (b) it licenses the alleviation of medicines and their attendant awkward results. The advantage of surgery shows up maintained for about 4 years even though some problems look like accruing. Some research studies have confirmed a much healthier standard of living of STN DBS. The very best potential candidates with regards to DBS are the ones by using a small period related to gain from specific dosages related to levodopa, individuals who have a considerable motor gain from oral thrush drugs, and people who might be restricted merely by dopa-induced negative effects. Mental issues, for instance, disorientation or perhaps even memory shortfalls, might be intensified by DBS, and it is a good relative contraindication regarding the processes [3].

Surprisingly, STN DBS might wound particular elements related to cognitive development. Activation configuration enhanced for motor benefit might impair spatial slowed down recall or perhaps even reactions inhibition. Others have discovered STN DBS might boost some executive features, but GPi DBS might create harmful results. Nonetheless, quite effortless cognitive duties might be unchanged or improved by STN DBS, whereas more difficult demanding tasks could be flawed. Socially essential actions such as the detection of the psychological mood of some irritating face might be flawed. STN DBS additionally might create awkward psychological results, such as frenzied results, hallucinations, lessened attitude and yet over at in some cases might offer an anti-depressant outcome [3].

5.2.4.4 DEEP BRAIN STIMULATION (DBS) WITH REGARDS TO DYSTONIA

Having the rise of DBS with regards to remedying PD and tremor, there arises an all-natural impulse to try it out really for dystonia. Stereotaxic ablations considering the GPi or perhaps even thalamus were utilized for a few years within the remedy for scientifically refractory comprehensive dystonia; nonetheless, their overall performance was not tremendous. Earlier results related to DBS with regards to dystonia concerned the thalamus overall and, of course, the GPi internal segment. Having the improving results related to pallidotomy with regards to comprehensive dystonia attributable to overall the DYT1 change, overall the GPi had become the main goal with regards to main dystonia, however, the thalamic objective remains utilized. Within the latest monitored experiment related to pallidal DBS about 22 affected individuals with main general dystonia, there arise a good 30%–50% enhancement of signs and symptoms. Unrestrained tests have even generated potentially promising outcomes with regards to main general dystonia. The smaller sized sequence of case research results have proposed possibilities usefulness with regards to managing main cervical dystonia plus some kinds of supplementary dystonia. While DBS with regards to managing dystonia needs more examination, earlier outcomes are potentially promising [4].

5.2.4.5   DEEP BRAIN STIMULATION (DBS) FOR TOURETTE SYNDROME

There have also been several of the latest results related to DBS with regards to Tourette Illness. The centromedian-parafascicular difficult considering the thalamus has been specified bilaterally within the majority of all those instances; however, the GPi and of course the frontal limb considering the inside capsule even have remained specified. To this point, six cases related to DBS with regards to Tourette syndrome have already been revealed, and there are inadequate records to check usefulness all over directs. Nonetheless, every affected individual has a level of tic decline along with DBS during these direct [5].

5.2.4.6   DEEP BRAIN STIMULATION (DBS) WITH REGARDS TO PAIN

DBS has been applied for longer than half a century as a treatment for various intractable aches signs and symptoms, such as neuropathic pain, phantom limb pain, failed lower back pain, as well as cluster-headache pain. Various written documents based upon anecdotal encounter or open-label research studies advise DBS offers short-or perhaps even lasting profit within a number of each of these symptoms. The advantage differentiates based on the duration of checking out, the condition of the property handled, the meaning related to enough pain alleviation, and of course, the site of activation. Sites related to activation include assorted beginning with the sensory thalamus towards the periaqueductal gray, periventricular gray, rear-end hypothalamus, internal capsule, and of course, the motor cortex. Some think that activation considering the periaqueductal gray or periventricular gray is especially suitable with regards to nociceptive pain, but DBS considering the sensory thalamus is significantly more useful for deafferentation pain. Research concerning six patients along with group headaches advised that DBS considering the ipsilateral anteroposterior hypothalamus gland lowers group headache assaults, however perhaps one of the affected individuals passed away coming from a presurgical intracerebral hemorrhage. The danger is certainly not harmless. Much higher elements related to activation, for instance, cortical direct might be very likely to cut down discomfort about post-stroke discomfort symptoms based upon open-label results. In addition, DBS considering the thalamus might cut down discomfort concerning phantom-limb illness based upon open-label analysis. Surprisingly one particular research study utilized useful magnetic resonance imaging (fMRI) to recognize surge within the rear end inferior hypothalamus about those with facial pain associated with short-lasting unilateral neuralgiform headache strikes along with conjunctival infusion as well as tearing and after that specified DBS in which area in which to supply pain alleviation for anyone patients. Monitoring evoked results to effectively painful causes could be a method to determine nociceptive receptors within the brain that might become relevant directs for getting a site related to DBS to alleviate that in fact kinds of discomfort. Also, nearby arena possible results linked to pain as well as reported back then related to operations might forecast stimulation variant that gets rid of discomfort (low frequency relieved pain; greater than 50 Hz) [5].

5.2.4.7   DEEP BRAIN STIMULATION (DBS) FOR DEPRESSION AS WELL AS COMPULSIVE CONDITION (OCD)

While the research studies are presently restricted, DBS might sooner or later use a task within the remedy for refractory melancholy. Research conducted recently discovered that DBS considering the subgenual cingulate white matter (WM) enhanced temperament concerning four of six people having treatment-resistant melancholy. Overall the medical investigators specified all of this zone due to the fact they possessed formerly confirmed elevated fluorodeoxyglucose (FDG) uptake gauged along with positron emission tomography (PET) in this region about those with depression. A single-issue document advised activation considering the mediocre thalamic peduncle additionally might get rid of depressive signs and symptoms. A few have advised the development of high quality related to life created by STN DBS in patients along with PD is principally a mirrored image considering the alleviation of depression rather than the enhancement of motor signs and symptoms [5].

DBS of the mutual frontal limbs considering the inside capsules might cut down signs and symptoms about OCD like present in three affected individuals in a single research study. An additional small, short-period, blinded research study confirmed that a couple of four affected individuals along with OCD possessed alternatively great or perhaps even low profit right after activation of one's frontal limb considering the inside capsule. Some kind of open-label research study located development related to OCD concerning three of four affected individuals. Activation considering the ventral caudate nucleus pleased depressive as well as OCD signs and symptoms in an open-label case document associated with a single affected person [5].

5.3 PROPOSED METHOD

5.3.1 MICROCONTROLLER NRF51822

Nordic semiconductor, nRF51822 Bluetooth low energy, 2.4 GHz wireless SoC offers a multiprotocol SoC suited for Bluetooth low energy and 2.4 GHz ultralow-power wireless applications. Nordic semiconductor, nRF51822 Bluetooth low energy, 2.4 GHz wireless SoC uses a 32-bit ARM^R Cortex^{TM} M0CPU with 256 kB flash and 16 Kb RAM [6].

5.3.2 BLOCK DIAGRAM

The stitcher is a microcontroller-based device which will generate pulse train, which in turn could be fed to required parts of the brain, to help their connecting or stitching together. The basic functioning of a stitcher chip can be explained referring to the timing diagram. In the various pins p1, p2, ..., pn of the chip, pulse trains of voltage v is made available in a sequential manner. In other words, the pin p1 will be having a pulse train of voltage Pv, for a duration of Pd to start with. Once this duration Pd is over there shall be a gap Pg of time before the pulse train appears on pin p2.

The sequence continues on other pins. Once the pulse train appears on all other pins one after the other (one round is completed), the same thing repeats from pin p1 for a second round. As many rounds are required or as much time as required, this process can be repeated. Since the chip is programmable amplitude, frequency, and delay can be altered. Once these values are set, the pins could be connected to the desired point in the brain of the subject [21, 22]. Figures 5.5 and 5.6 shows the block diagram and flow diagram representations. SMART stands for self-involved motivated action reward technique. By combining modified ICSS and giving a suitable environment is envisaged, which will provide enough and more of involvement to the participating subject. This, in turn, will increase the neuronal density of the participants and hence enhance their capabilities. SMART hall is such an arrangement where the model method of rewarding is integrated with traditional methods of recreation available with a suitable environment.

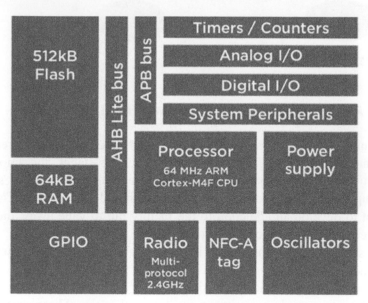

FIGURE 5.5   Block diagram.

Every traditional recreational facility is attached with an intracranial stimulation giving extreme pleasure to the participating subject. As a reward, intracranial stimulation very pleasing, the motivation will be very high, eliciting maximum self-involvement. Different modified ICSS facilities will be arranged in a big hall. Subjects can choose their facilities as they like. As each one goes to a particular facility, the user is identified by the RF id tag wears by them, and the system adjusts the response accordingly. If the user gets attached to a single facility, the system takes the steps to somehow repel that one from that facility [8]. This is done as follows: (1) Repelling the subject from the addiction; (2) Single Person SMART Hall; (3) Multiple Person SMART Halls; (4) Technical requirements of SMART Hall; and (5) Social nature of Multiple Person SMART Halls.

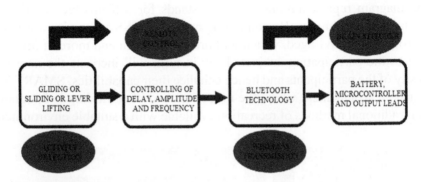

FIGURE 5.6   Flow diagram.

5.3.3   REPELLING THE SUBJECT FROM THE ADDICTION

As the number of the session goes up the no of times action to be repeated for stimulation goes up. Together with that in later stages, the amount of stimulation received also goes down. At a later stage, if the same subject comes up for the same facility, it gives a cold shoulder to the user. This will help not to develop an addiction to the facility but to vane away, the child from the facility is a due course of time.

The social nature of the facility is very helpful in many ways. There will be interaction and exchange of ideas among the participants. Thus it will not become a machine interactive session. As one does, the other can see and learn a lot. One can, in many cases, teach the other one. These aspects will increase the communication of learning and will automatically improve the capabilities of the brain [8].

Other than small and silly actions like pressing a pedal, the traditional recreational facilities in the rich cage are made into activities required to get stimulation in a smart hall. As the traditional activities require more involvement and the stimulation obtained is highly pleasing, both objectives will be met. When the subject completes one oscillation on an ordinary swing, it gets stimulation. Similarly, when a subject completes a sliding on an inclined plane, it gets stimulation. Later for n repetition of the same activity will get one stimulation. Here n can be range from 1 to 3 or 4. The activity to be done to get stimulation is changed, which is given with some audio or visual instruction, thus the hunting for pleasure will be intensified, and the brain will grow.

5.3.4 SINGLE PERSON SMART HALL

In the case of a single individual in a hall, different entertaining facilities could be set up. The subject shall wear an electronic kit and electrodes are implanted into his brain. As and when the subject does various activities, the connected system sends a pulse to the subject kit, which it translates to a proper stimulation. The pulse to be sent may be sent by Bluetooth (a wireless transceiver module) as the subject will have to be walking around without constraints [8].

5.3.5 MULTIPLE PERSON SMART HALLS

If multiple subjects are to be there in the SMART hall simultaneously, then there should be proper arrangements to identify each subject interacting with each system of recreation. For this RFID, technique (which includes a small range electronic tag with transceiver antenna and wireless technique support) could be used [22, 23].

5.3.6 TECHNICAL REQUIREMENTS OF SMART HALL

Thus the SMART hall should be in a technical manner. Various conventional recreational facilities should be provided. Arrangements should be there to identify that a particular subject has utilized the facility. Then the systems should be able to identify the status of that particular subject like a fresher on this facility or experienced one. Based on the level, the central system instructs the particular facility to send a pulse to the subject, which is to give a stimulation ranging from 0 to 1 in intensity. Depending on the experience level, proper audio/videos instructions should be given out. In the case of saturated subject information like no more good out of trying today or anymore on this should be given. The intelligence could be built into the central computer or in the individual chip, the subject carriers effortlessly, in an unreachable way to the subject. The chip could be embedded under the skin behind the ears or some other convenient place near the brain. Another subject who is just watching or trying to do a drill should not be getting stimulation, just because a nearby subject did the drill property. The pulse given to one subject is useful to that subject only. The structure of the modified ICSS experiment is shown in Figure 5.7.

The SMART hall proposed provides an environment where the subject will be highly motivated, highly involved, exploring thoroughly, getting appreciation and acknowledgment from peers, learning by seeing the peers performing, taking cues from viruses clues, efficiently learning by feedback, de-addicted, acquiring a social nature and so on. The subject can become smarter faster by using SMART hall [12].

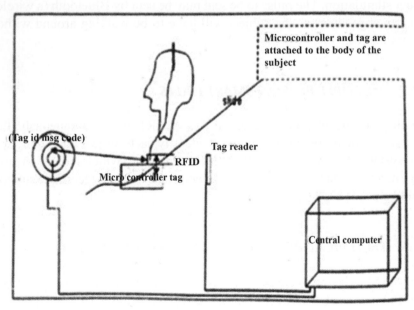

FIGURE 5.7   Modified ICSS experiment.

5.3.7   SOCIAL NATURE OF MULTIPLE PERSON SMART HALLS

As many subjects are working in the SMART hall, they can see each other doing and the one who is doing it well can be proud of doing so in front of many. This works like an acknowledgment and appreciation for their performance. When they played for their Institutes and won laurels, it made an impact on them. They felt important, wasted to grow further, and so on. Thus the component of social acceptance and appreciation and so on are to be incorporated in efforts to improve the capabilities of mentally challenged children. By proposing a SMART hall for multiple people, this thesis is aiming at a very efficient method for the same [8].

5.4 IMPLEMENTATION AND RESULTS

Intracranial self-stimulation (ICSS) can be considered an archetype wherein rodents self-administer gratifying electrical activation (also known as cerebrum activation incentive (BSR)) via electrodes fixed directly into the cerebrum. Although the phrases are sometimes utilized interchangeably, ICSS will be the behavior and mood as well as BSR, which happens to be obtained from the behavior and mood. The permutations of the ICSS paradigm are as follows: monopolar or bipolar stimulation takes place, rats perform self-administrator stimulation in the brain, and numerous approaches are used [11].

By using a standard operant chamber, the testing process takes place. Make a hole in the center part of the top portion of the chamber. Groove is made within the front side of a given space to effectively enable managing overall the creatures before the following checking. A ring-stand, as well as customary clamp, is utilized to droop overall the commutator beyond the middle considering the hollow. Overall the operant space might be positioned inside-outside cabinetry along with doorways that allow checking below light-as well as sound-controlled circumstances, even though this is not crucial due to the reason that the gratifying usefulness of MFB activation results in behavior and mood extremely thorough [12].

Manipulations that raise incentives and such that often raise sensitivity create unlike behavioral outputs about ICSS assessments. ICSS is likewise worthwhile since it might be employed to learn the period path of experimental manipulations, for instance, intense medication results or perhaps even persistent modifications in gene developments. Nonetheless, it isn't especially helpful for about research studies that can determine procurement charges, because lab rats usually choose the operant in the context of a moment when the activation electrode is correctly positioned. BSR might be served via alternatively monopolar or sometimes-bipolar electrodes. The utilization of monopolar electrodes is a lot more than related to bipolar electrodes, due to its smaller sized length. Overall the cathodal rate is utilized considerably more than related to anodal rate. Additionally, it is more dependable compared to anodal rate, which could harm cell tissues merely by emanating metallic ions beginning with the electrode; actually, some kind of anodal rate is most often utilized by the end related to ICSS with the electrode tip. Therefore, we change the frequency (Hz) as opposed to the magnitude (A) considering the activation because we wish to continue to keep regular the people related to neurons that are triggered. Generally, so we analyze animals within a sequence of fifteen 1-min tests, almost every on a varying activation frequency. Each 1-min tryout includes a simple 5-s time frame in which the animal obtains completely free priming trains (1 s1) considering the cathodal (monopolar) activation that is accessible, leading to a good 50-s time frame in which the range of results (switch pushes with regards to lab rats) is considered, leading to a good 5-s time-out time frame through this the frequency is lessened within a 0.05-log unit action [12].

5.4.1 MATERIALS REQUIRED

The material requirements are as follows:

➢ Rats (in theory, any kind of strain related to lab rats may be used): Careful experimenters need to comply with nationwide as well as department recommendations regarding the concern and utilize related to research centers creatures, such as regional specifications with regards to surgical procedure anesthetic. Crucial utilization of adult creatures (lab rats = 300 grams) is usually recommended considering that electrodes that might be completely added on towards the skull might be vagrant merely by bone development).
➢ Dental cement that is used: acrylic for lab rats.

The equipments used are as follows:

• Stereotaxic instrument.
• Acrylic cage with pneumatic actuators as pedal.
• Brain stitcher.
• Wireless connector.
• Computer and software to automate ICSS.
• Digital oscilloscope.

- Surgical procedure products (anesthetic; atropine; antibiotic; antibiotic cream; syringes; equipment (scalpel, hemostats); scalpel blades; sutures; based on overall the organization, sterile products as well as post-operative medications might be necessary).
- Electrodes (monopolar).
- Electrode possessor with regards to the stereotaxic instrument.
- 1/8-in drill bit.
- Stainless microscrews.
- Adaptable cable leads.
- Software.

Electrodes (monopolar) ought to be covered along with PFA insulating material excluding with the tips, which happens to be trim upright along instead of pointed. Overall the anode ought to be some kind of uninsulated adaptable stainless line. Commercially accessible electrodes are sometimes developed in a way that likewise covered electrode (0.25 millimeters) and of course, the uninsulated anode (0.125 millimeters) are inserted within the text along with thread connect, which inserts by using a screw-down adaptable line bring. Using this product or service, all of us divide electrode by itself (without the along with thread connect) to the duration of 10 millimeters with regards to lab rats. The electrode used is shown in Figure 5.8.

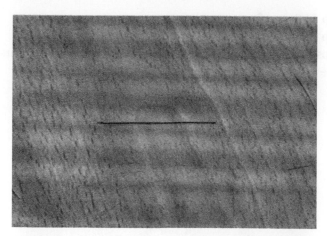

FIGURE 5.8 Electrode.

- Device set up: There are no exact specifications regarding the layout (measurement, resources) considering the operant training compartments, while the manipulandum (switch with regards to lab rats)ought to be metallic and certainly attached reduced using one surface (3 centimeters beyond the ground to obtain lab rats) to enable remarkable retorting charges. Overall the front side surface of the space ought to be see-through make it possible for visible assessment of a given subject. Overall the compartments might be positioned inside sound-as well as light-proof outside cabinetry (along with peep-hole make it possible for visible assessment considering the creatures); this is not essential to guarantee trustworthy overall performance, even though regular method (no more outside cabinetry, outside cabinetry doorways consistently wide-open or perhaps even outside cabinetry doorways, consistently shut down) should be implemented. The trimmer used is shown in Figure 5.9. The proposed brain stitcher is shown in Figure 5.10. The overall cage and dilling machine are shown in Figures 5.11 and 5.12.

- Overall, the stimulators need to have the possibility to provide square-wave activation over the series of ~10–400 μA, and also have a differentiate results observe by which some kind of oscilloscope is included.

Procedures include:

- **Step 1:** Precursory procedures.
- **Step 2:** Surgery.
- **Steps 3–5:** Preliminary training programs as well as the persistence of lowest valuable current steps.

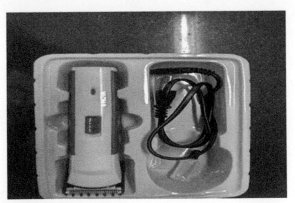

FIGURE 5.9   Trimmer.

FIGURE 5.10 Brain stitcher.

FIGURE 5.11 Cage.

FIGURE 5.12   Drilling machine.

- **Steps 6–8:** Rate-frequency setting.
- **Step 9:** Checking.
- **Step 10:** Verification.

1. Starting guidelines subsequent emergence considering the creatures towards the group, abandon them to be able to adjust. Acclimation related to 3–7 d is usually recommended.
2. Stereotaxic arranges related to electrode positions, the total number of stainless-steel screws fixed directly into the skull.
    a. Anaesthetize almost every rat (≥300 grams) based on department specifications, as well as can impair it really within a stereotaxic apparatus.
    b. Insert a monopolar electrode directly into MFB with the level of overall the lateral hypothalamus (LH) (2.8 mm posterior to bregma, 1.7 mm lateral to the midline, 7.8 mm beneath Dura.
    c. Should wrap overall the anode about a couple of considering the four stainless screws that might be along with thread directly into the skull to hold down overall the electrode assembly.
    d. Design the electrode assembly merely by masking overall the screws and of course, the electrode along with acrylic oral cement.
    e. Close down about boundaries of one's incision along with sutures in a way that the skin layer wraps securely all over (instead of above) so far the electrode assembly.
    f. Go through the cut along with the external triple antibiotic gel. Apply postsurgical medications based on department restrictions.
3. Preliminary training programs, as well as determination related to the lowest useful current. Let the subject to get better with regards to one week without the need for checking.

4. During this period, watch restoration following department specifications. Right after one-week restoration, put the subject within the operant training compartments as well as practice all of them on any fixed-ratio-1 plan (FR1) to get back to for cerebrum activation like explained.

5. Every time overall the rat pushes overall the switch, the pc will need to provide a 0.5-strain of square-wave cathodal impulses (0.1-ms pulse period) at a put frequency related to 141 Hz. After supply considering the activation, make use of a 0.5-s breakthrough this even more results typically are not strengthened. One particular day-to-day training session related to 60–90 min is advisable 5 d a week to accustom overall the subject towards the actual physical needs considering the ICSS processes. Overall the activation current (~100–300 μA to obtain lab rats) ought to be modified via the investigator towards the most affordable value that helps a good trustworthy value related to retorting (≥40 incentives for each minimum) for three successive days. This can be deemed overall the 'minimal current,' reflecting sensitiveness towards the gratifying outcome of overall the activation. As soon as the minimum current is noted for any animal, it can be carried out regular regarding the remainder of the research. Figure 5.13 shows the surgery process.

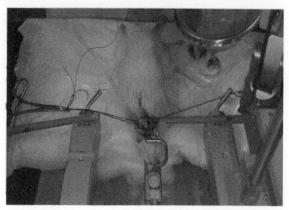

FIGURE 5.13   Surgery.

6. Adjust each animal to short assessments by using a climbing down the sequence of 15 activation frequencies with the lowest useful current.

7. Immediately after retorting has been examined every of one's 15 frequencies, replicate this treatment in a way that each animal

receives six these types of series each day (90 min of conditioning). In the course of the training programs processes, the various frequencies ought to be modified for any animal to ensure the top-notch 6–7 frequencies continue responding; this method allows equivalent sensitiveness to treatment options that raise or perhaps even decrease ICSS thresholds.

8. Calculate ICSS thresholds (the regularity of which overall the activation turns into gratifying) merely by plotting a graph [1].

5.4.2   RESULT ANALYSIS

After fixing the product, we stimulated the rats for 15 minutes twice a day during a week and continuing for two months. The verifications will be obtained after two months using the process called staining. By varying the frequency and current, we found out that the maximum performance of pedal pressing will be delivered at 140 Hz with 300 uA. The following graphs are the statistical analysis by varying parameters. Figure 5.14 shows the representation of low current, Figure 5.15 shows medium current representations, Figure 5.16 shows the high current representation, Figure 5.17 with 120 Hz, Figure 5.18 with 140 Hz, and Figure 5.19 with 160 Hz. Table 5.1 shows the comparison of graphs concerning its current and rate of pressing.

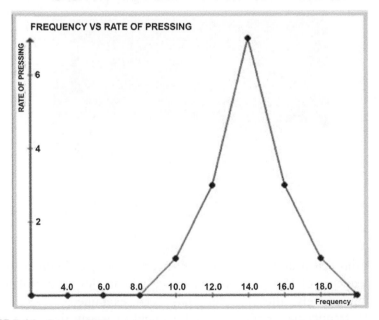

FIGURE 5.14   Low current.

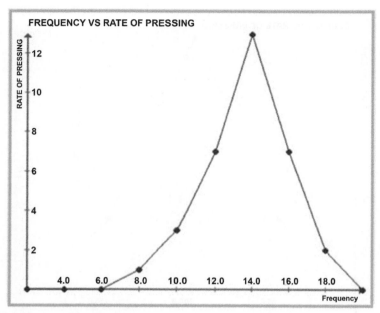

FIGURE 5.15   Medium current.

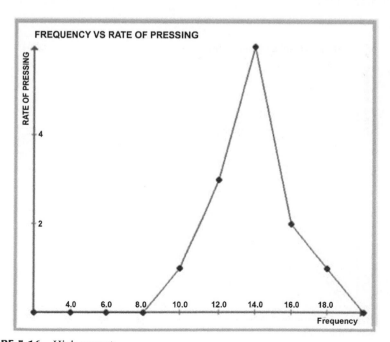

FIGURE 5.16   High current.

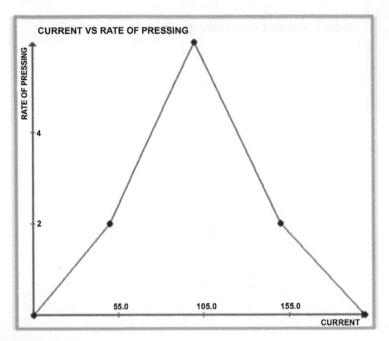

FIGURE 5.17   120 Hz.

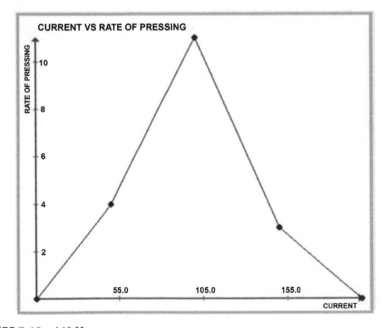

FIGURE 5.18   140 Hz.

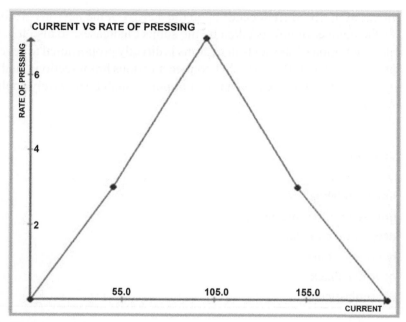

FIGURE 5.19   160 Hz.

TABLE 5.1   Comparison of Current vs. Rate of Pressing with Various Frequencies

Frequency	Current	Rate of Pressing
17.120 Hz	1650	6
18.140 Hz	1750	8
19.160 Hz	2050	11

5.5 CONCLUSION AND FUTURE SCOPE

This product helps in including the million especially differently-abled humans around the world into the path of evolution and to help them create for themselves a platform where they can involve in contributing to the development of the society with the help of this chip. This chip helps them in stimulating their brain activities to an extent that is considered desirable to boost their performance and be a productive human resource. This will be a promising product that can be considered a cure for various diseases that are closely about brain activity. They included conditions like epilepsy, Parkinson's disease, chronic pain, dystonia, and even for normal people.

Efficacy of this product depends upon the density of the neurons prolifer-ated. As the number of efforts taken by the subject increases, so also it helps the brain. Formation of shortcuts in neurons is directly proportional to several attempts of the subject. Brain stitcher connects various brain sections, and by this brain stitcher method desired amount of pulses shall be precisely applied at specified durations.

KEYWORDS

- **brain modeling**
- **brain-computer interfaces**
- **deep brain stimulation**
- **neurocontrollers**
- **neurofeedback**
- **neuroplasticity**

REFERENCES

1. Carlezon, W. A., & Chartoff, E. H., (2007). Intracranial self-stimulation (ICSS) in rodents to study the neurobiology of motivation. *Nature Protocols*, *2*(11), 2987–2995. doi: 10.1038/nprot.2007.441.
2. Alpaugh, M., et al., (2019). A novel wireless brain stimulation device for long-term use in freely moving mice. *Scientific Reports*, *9*(1), 6444. doi: 10.1038/s41598-019-42910-7.
3. Eric, R. K., James, H. S., & Thomas, M. J., (2000). *Principles of Neural Science* (4th edn.). London: McGraw-Hill Health Profession Division.
4. Melo-Thomas, L., et al., (2017). A wireless, bidirectional interface for *in vivo* recording and stimulation of neural activity in freely behaving rats. *Journal of Visualized Experiments: JoVE, 129*. doi: 10.3791/56299.
5. Tandon, S., Kambi, N., Mohammed, H., & Jain, N., (2013). Complete reorganization of the motor cortex of adult rats following long-term spinal cord injuries. *The European Journal of Neuroscience*, *38*(2), 2271–2279. doi: 10.1111/ejn.12218.
6. Pinnell, R. C., Dempster, J., & Pratt, J., (2015). Miniature wireless recording and stimulation system for rodent behavioral testing. *Journal of Neural Engineering*, *12*(6), 66015. doi: 10.1088/1741–2560/12/6/066015.
7. Halpern, C. H., Attiah, M. A., Tekriwal, A., & Baltuch, G. H., (2014). A step-wise approach to deep brain stimulation in mice. *Acta Neurochirurgica*, *156*(8), 1515–1521. doi: 10.1007/s00701-014-2062-4.
8. De, H. R., et al., (2012). Wireless implantable micro-stimulation device for high frequency bilateral deep brain stimulation in freely moving mice. *Journal of Neuroscience Methods*, *209*(1), 113–119. doi: 10.1016/j.jneumeth.2012.05.028.

9. Nordic Semiconductor. (2019). *NRF51822.* [Online]. Available at: www.nordicsemi. com (accessed on 18 December 2020).

10. Vithayathil, M., Jarin, T., Prabhu, S. R. B., Ningthoujam, R., & Surajkumar, S. L., (2019). Designing and modeling of a low-cost wireless telemetry system for deep brain stimulation Studies. *Indian Journal of Science and Technology, 12*(8), 1–13. doi: 10.17485/ijst/2019/v12i8/141815.

11. Vithayathil, M., Athappilly, G., Prabhu, S. R. B., Paul, M., Ningthoujam, R., & Surajkumar, S. L., (2020). Neural proliferation using brain stimulation methods intended for pediatric neuropsychiatric population: A hypothesis and theoretical investigation. *TEST Engineering and Management, 82*(1), 9138–9151.

12. Vithayathil, M., Ningthoujam, R., & Surajkumar, S. L., (2019). *Embedded Based Solution for Intracortical and Intracranial Microstimulations for Assessing the Behavior of Rodents.* Bangalore: IEEE.

CHAPTER 6

INSIGHT INTO VARIOUS ALGORITHMS FOR MEDICAL IMAGE ANALYZES USING CONVOLUTIONAL NEURAL NETWORKS (DEEP LEARNING)

S. SUNDARESAN,[1] K. SURESH KUMAR,[2] V. KISHORE KUMAR,[2] and A. JAYAKUMAR[2]

[1]Assistant Professor, Department of Electronics and Communication Engineering, SRM TRP Engineering College, Trichy, Tamil Nadu, India, E-mail: sundaresanece91@gmail.com

[2]Associate Professor, Department of Electronics and Communication Engineering, IFET College of Engineering, Villupuram, Tamil Nadu, India, E-mails: sureshkumarskphd@gmail.com (K. S. Kumar), v.kishore882@gmail.com (V. K. Kumar), jayangce@gmail.com (B. J. Kumar)

ABSTRACT

Deep learning (DL) is occupying a crucial part in the applications related to computer vision that might show various remarkable performances throughout the applications. The major obstacle identified in the DL applications is its complexity in computing. So, there needs an algorithm to reduce the computational complexity in its applications. Many diseases are diagnosed from the images taken in CT and MRI scans. This chapter aimed at the early prediction of Alzheimer's disease (AD), an algorithm that discriminates the mild cognitive impairment (MCI) and cognitive normal (CN) is casted-off, which shows better results in its analysis. Another survival model is used for the improvement in the accuracy of the prediction, followed by the reduction of ring artifacts in the images. Three consecutive methods are implemented one after the other, which generates a corrected image by means of correlation. In another method for the identification of tuberculosis (TB), a CNN mixed with LSTM is utilized. This shows better results and reduces the computational complexity to the core maximum. This is achieved by retrieving the microscopic images collected from infected persons. CNN is trained in such a way to achieve a prediction accuracy of 99.99%. For estimating the feature of the images, a CNN that is trained earlier is used along with a support vector machine (SVM) classifier and the multilayer perceptron (MLP). For establishing the stable learning framework, a deep CNN is presented along with a cascaded structure that will increase the accuracy. On the whole, this chapter describes some of the methods for reducing the computational complexity of the medical image classifications, thus increasing the chance of disease prediction.

6.1   INTRODUCTION

In clinical practice, clinical imaging plays an undeniably significant job in educating the dynamic of clinicians for malady the executives. The exceptional development in PC vision lets to utilize an immense measure of clinical picture information for the analysis, treatment, and itemized report of the illnesses. By removing the highlights like shape, surface, shading, and earlier information from the obtained clinical pictures information, present-day PC vision advances like AI, picture division, design arrangement, and so on empowers a productive method for illness analysis than research facility specialists. Since pictures are given in a 2D group containing pixel esteems, it is simpler to utilize fundamental picture handling systems, which will be a lift for proficient infection determination and treatment utilizing PC vision techniques. Be that as it may, it is consistently a major test in overseeing PC vision in clinical imaging because of the multifaceted nature in the clinical pictures. Radiomics is a precise way to deal with study the idle data in clinical imaging for showing improvement about the exactness in guess. A common radionics study includes picture procurement, highlight extraction, including examination and prescient displaying for a clinical result, for example, quiet endurance [11]. Endeavors ought to be present in order to normalize computable imaging highlights (radiomic highlights) by actualizing libraries available in open source, for example, PyRadiomics [12]. The element store consists of a huge number of human-created recipes, intended on the way to extricate the circulation or surface data from clinical pictures. In radiomics considers, an element decrease technique (e.g., rule part investigation) is utilized to select delegate highlights [13]. The prognostic highlights are normally decided utilizing Cox proportional hazard model (CPH) [14]. In the previous years, a few radiomics highlights might have indicated the extrapolative incentive in various infections, particularly various kinds of malignant growth [15]. In any case, the maximum measurable atmosphere of radiomics highlights will make the component determination inclined towards numerous analyses, prompting bogus positive and minimum level of execution happening with the approval partners.

Alzheimer's disease (AD) is a dynamic dementia, impact seniors that outcomes in losing of association between nerve cells. Attributable to Alzheimer's ailment mind get recoils, hippocampal size gets diminished and broadened cerebrum ventricles. As Alzheimer's sickness advances, it spoils memory, thinking capacity, and face issues in day-to-day action. By knowing about the Alzheimer's malady, MCI, and CN indication is a unique model which is used for utmost testing undertaking that nervous system specialist appearance on or after recent existences. Promotion is analyzed utilizing the tests done physically and mini-mental state examination (MMSE) [16, 17]. As a clinical system, image creates the neural imaging system that assumes a significant job for analyzing the auxiliary and changes made practically in the cerebrum to incorporate. Various imaging schemes are used, and each might have certain properties that enhance the imaging capacity. For each part of the human body, separate imaging schemes are used. Inferable from basic access, MRI is utilized to separate assistant changes caused in light of AD, CN, and MCI sign. The most notable MRI groupings are T1-weighted and T2-weighted yields. T2-Weighted yields are presented. As showed up in Figure 6.1, it is seen that AD MRI having cortical rot, change in hippocampal size and expansion of ventricles appear differently in relation to CN and MCI with clarification. It is confirmation that the surface of the brain get changes from CN to MCI to AD. The T2w picture had an inclined scattering for both GM (grey matter) and WM (white matter) with long tails demonstrating lower homogeneity and discriminability for these tissues. Thus, T2w had non-front of power regards between WM and GM, which gives better discriminability between tissues. Hence, T2 weighted MRI castoff for tissue portrayal for assessment of neurological issue.

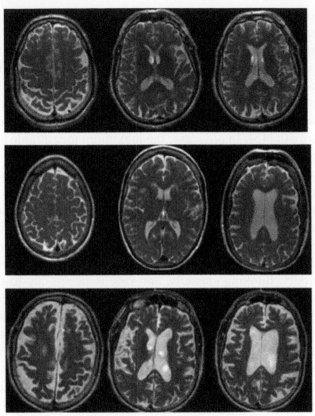

FIGURE 6.1   MRI images of CN, MCI, and AD cross-sections.

Microscopy innovation of utilizing magnifying lens to see zones of interests, which can't be seen by unaided eye is for the most part considered as hotspot for clinical imaging information. Regular optical or light microscopy includes transmitting obvious light through and reflected back from the specimen through the focal point to see the example amplified. The resultant picture can be straightforwardly observed by the eye or can be carefully caught. In low-asset illness inclined regions, microscopy is usually utilized for symptomatic undertakings, in view of being both basic and adaptable. New advances like, analysis dependent on sub-atomic science and stream cytometry can't be managed in such places. Regardless of whether magnifying lens is accessible, because of the lack of gifted experts, sickness finding regularly rely on illness side effects and clinical signs alone, which is inclined to mistake and results in death, medicate obstruction and loss of cash for purchasing superfluous medications. In this way, there ought to be another proficient option for quality analysis that isn't accessible today in numerous spots.

Development of deep learning (DL) techniques brings about tremendous progression in the field of PC vision to the degree that robotized object acknowledgment with precision beating human ability [18]. Numerous past issues rely close by designed picture highlights for a specific assignment, which are utilized by AI calculations. Instead of depending on this, there ought to be a typical methodology in clinical imaging, which can learn valuable picture portrayals consequently with satisfactory layers in the model. Thusly, this chapter especially centers on the use of profound figuring out how to perform microscopy-based purpose of care analysis for tuberculosis (TB). The pictures are caught by affordable gadget comprises of advanced cell joined magnifying lens, which is proposed to use in an asset obliged condition. Additionally, the proposed model plans to decrease the computational intricacy by lessening the quantity of learnable parameters in the model.

6.2 ALZHEIMER'S DISEASE (AD) CLASSIFICATION USING CNN

The recent development of the neuroimaging in association with DL motivates the possibilities of enhancing the early identification of AD at the earliest. The physicians are finding it easy for diagnosing the diseases using machine learning. This could be possible because of MCI (mild cognitive impairment) and CN (cognitive normal), as the AD is predicted from this has advanced to worldwide. The identification of AD is done accurately through the differentiation of CN and MCI. In this, multiclass classification and binary class classification are cast-off with total 4463 slides in which some are separated for training and some for testing. Totally two set of MRI images are taken named as T1 and T2 training sets. The slices of the MRI images from CN, MCI, and AD images are given in Figure 6.1. By doing so an accuracy, sensitivity, and specificity are obtained as 99.99%, 99.99%, and 99.99%, respectively for AD-CN. In the case of AD-MCI, it is obtained as 96.2%, 93%, and 100%, respectively. For CN-MCI, it is obtained as 98.0%, 96%, and 100%, respectively, and for AD-MCI-CN, it is obtained as 86.7%, 89.6%, and 86.61%, respectively. This could be tested after using a ten-ranged validation which is of 10 cross fold and by this, an accuracy of 98.0% is obtained. The results are generated by improving the extrapolation of AD from the CN and MCI combination, which is then compared to the flow of work continuously and is then used for differentiating the AD at the beginning stage. In this work, aggregate volumes of MRI images with 137 nos. are acquired from the Alzheimer's disease neuroimaging initiative (ADNI). The collected subjects are totally of 120 and of these 70, 25, 20-AD, MCI, and CN are collected, respectively [1]. Table 6.1 shows the MRI image data set representation from the ADNI data set.

TABLE 6.1   MRI Image Data Set Representation from ADNI Data Set

Disease Types	Total Samples	Score of MMSE	Total Score of GD Scale	Age in Years	Total Volume of MRI Data's	NPI-Q Total Score
AD	70	30–35	0–25	>60	60	0–35
CN	32	36–40	0–25	>35	40	0–35
MCI	40	23–27	0–25	>70	45	0–20

This work has analyzed the enactment of the classical model on two weighted MRI images, data sets that are acquired from the subjects which are of the same data set, and this is evaluated from the different parameters that are under the multiclass classification and the binary class. For performing the segmentation, MRI images are sliced initially into a certain size and are sending to the preprocessing unit. This is for correcting the geometric distortion, thus by reducing the noise in the Gaussian filter. The architecture of the new novel model is described in Figure 6.2.

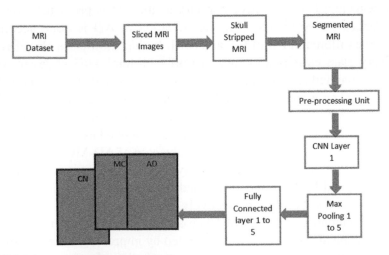

FIGURE 6.2   Architecture of modified system.

6.2.1 SKULL STRIPPED ALGORITHM

By using the skull stripped algorithm, the non-brain soft tissues remained detached from the slides. The changes that occurred on over the brain cells by the surface in addition to structural models might remain kept for differentiating the strong person and the diseased person. A modified ICA is used for performing the brain pathology segmentation. Here the AD disease is classified and established on GM. It might have neuron cells and non-neuron cells. This GM potency goes under development and ensures for growth of the child into adolescence. This is made to carry out the glucose levels to the brain, and the changes in the brain nerves may create changes and affects the memory. The main focus is for classifying the AD from gray matter. The CNN might use a classifier that is somehow different from the computer visualization methods. The classification methodology might contain three major parts. The Pre-processing task, testing, and training the model, finally test by using the 10 fold cross-validation techniques. Skull stripping algorithm is an earlier processing method [2]. For the fine-tuned image classification non-neurons and the annoying part of brain cell tissues might be removed from slides at the earliest. The noise and details of the information is reduced by using the Gaussian filter that is applied just before the initial stage of skull stripping. The steps in the algorithm are described below:

➢ **Algorithm:**
- **Step 1:** Let $I_i(X, Y)$ is considered as the input image from the data set, $K(X, Y)$ is the kernel data set and I_o is considered as the obtained output image.
- **Step 2:** Let I' is considered as the enhanced image.
- **Step 3:** Perform convolution process of input image $I_i(X,Y)$ and kernel $K(X,Y)$.
- **Step 4:** Thresholding the images by using certain methods to limit up to 254 ranges.
- **Step 5:** Enhancing the image I_1 to I_2.
- **Step 6:** Active contour by performing the active contour on I_3 to form I_4.
- **Step 7:** Output image: $I_{out} = I(X,Y) \times I_4$.

6.2.2 SEGMENTATION ALGORITHM

The concept of the proposed algorithm is Hybrid Enhanced ICA, and it is processed by introducing the concept and is mutually classified into the classes. The tissues of the brain in MRI image Data set is done by segmentation-based approach. The EM algorithm and K-means algorithm are used in a combined manner and is made as a hybrid algorithm for dealing with the brain tissues in datasets obtained from MRI. For some of the calculations related to mean and variance is obtained by using a k-means algorithm. Normally for this approach, the expected maximization algorithm is utilized [3]:

- **Step 1:** $G(X, Y)$ is considered as the input images that are divided into K identically self-regulating GMMs.
- **Step 2:** $G(X, Y)$ is represented into $G_i\{$, where $i = 1, 2, 3,...N$) vector.
- **Step 3:** The mean and the variance is calculated from the Gaussian mixture model using modified k-means algorithm.
- **Step 4:** The N pixels are partitioned into k sets of equal size.
- **Step 5:** The centroid is calculated.
- **Step 6:** Calculate the distance between the cluster centers and the Euclidean distance.
- **Step 7:** Determination of centroid in a particular $G(X,Y)$.
- **Step 8:** Calculate the clusters of the centroids once again.
- **Step 9:** Repeat the steps 2 to step 8.
- **Step 10:** On the off chance that the separation among $g(x,y)$ and different bunch community is not exactly or equivalent to the past separation at that point $g(x,y)$ might be in a similar bunch else it move to another group dependent on the separation.
- **Step 11:** The procedure get proceed till the bunches get combination.
- **Step 12:** The mean and the variance is collected.
- **Step 13:** The probability is been calculated.
- **Step 14:** If the difference between the old probability and the new probability is equal to zero, then the process stops.

The CNN performance depends upon the weights and architecture of the network. CNN depends upon the task, which is specific, and the data required for the network is already identified. The convolution layer dimensions are then modified into different stages like 3×3, 4×4, 5×5, and 3×3. The max-pooling layers might be useful for the extraction of topographies, and associated layers are made towards the count of 6. The backpropagation algorithm is used for the training purpose. Table 6.2 presents the set sizes, validation set, and the training set followed by Tables 6.3 and 6.4 show the performance parameters of MRI images for trained set 1 and set 2, respectively. Figures

6.3 and 6.4 display the graph for performance parameters of MRI images with respect to the trained set 1 and set 2, respectively.

TABLE 6.2 The Set Sizes, Validation Set, and the Training Set

Various Types of Classifications	Class Label	Training	Test Set	Total Images
Multiclass classification	AD	2137	564	2701
	MCI	887	223	1110
	CN	980	321	1301
Binary class classification	AD-MCI	2779	756	3535
	AD-CN	3224	798	4022
	CN-MCI	2314	543	2857

TABLE 6.3 The MRI Images of Trained Set 1 (Performance Parameters)

Type of Classification	Modeling	Accuracy	Sensitivity	Specificity	AUC	Images Used for Test	
Binary Classification	AD-CN	56%	84%	68%	90%	AD	486
						CN	241
	AD-MCI	54%	63%	82%	72%	AD	486
						MCI	212
	CN-MCI	23%	34%	22%	21%	CN	243
						MCI	223
Multiclass Classification	AD-CN-MCI	38%	46%	32%	67%	CN	243
						MCI	209
						AD	486

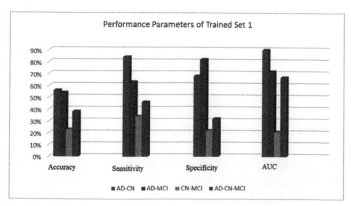

FIGURE 6.3 Graph for the performance parameters of trained set 1.

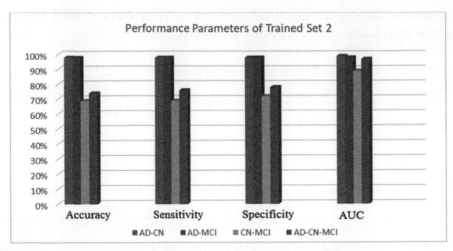

FIGURE 6.4   Graph for the performance parameters of trained set 2.

TABLE 6.4   MRI Images of Trained Set 2 (Performance Parameters)

Classification Type	Modeling	Accuracy	Sensitivity	Specificity	AUC	Images Used for Test	
Binary Classification	AD-CN	98%	98%	98%	99%	AD	486
						CN	241
	AD-MCI	98%	98%	98%	98%	AD	486
						MCI	212
	CN-MCI	69%	69%	72%	89%	CN	243
						MCI	223
Multiclass Classification	AD-CN-MCI	74%	76%	78%	97%	CN	243
						MCI	209
						AD	486

6.3   CNN-BASED ALGORITHM FOR ARTIFACT REDUCTION

Normally the quality of the reconstructed images taken from CT images might degrade because of ring artifacts. So, a hybrid model is introduced for the reduction of such problems by using CNN, as it processes some corrected images by fusing the image domain and sinogram domain images.

Initially, for doing the process, a training CNN is recognized that contains domain corrected images with certain artifacts (e.g., the ring artifacts). Next, the original images and sinogram domain amended images are given as input to the CNN trained image set for generating the images with fewer artifacts [4]. At last, the mutual correlation of image is fed to generate a corrected image by forming a hybrid nature via combining the data's from the ring artifacts, thus reducing the domain of sinogram and output using CNN. The results show that the anticipated technique might get reduced by the ring artifacts, which are deprived of distorting the structures. At first, for the earlier correction mechanism, the Fourier filter method using Wavelet transform is casted-off [5]. The image dataset obtained from the CT image set consists of artifact-free, pre-corrected images and is recognized to train a CNN.

6.3.1 CNN DESIGN

The new proposed structure of the network is shown in Figure 6.5. This CNN network consists of convolutional layers along with input and output. The database (DB) in every case consists of an image dataset that could be combined to the ring artifact image and the image modified earlier. The small patches of image are located at the same location in the artifact-free image, and this is again considered as the CNN target at the time of training. Figure 6.6 displays artifact-free image, image modified earlier and the ring artifact image. There are three samples in the image data set and the rows in each case represent individual samples.

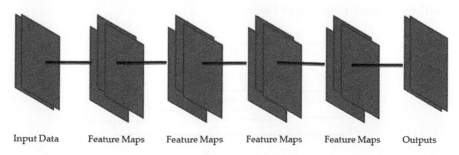

Input Data   Feature Maps   Feature Maps   Feature Maps   Feature Maps   Outputs

FIGURE 6.5   The proposed network structure.

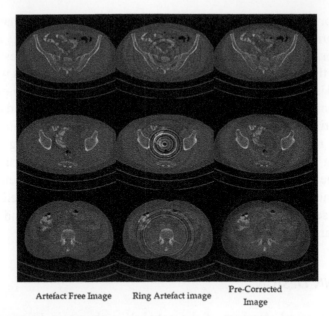

Artefact Free Image   Ring Artefact image   Pre-Corrected
                                               Image

FIGURE 6.6   Database samples.

In light of CNN, the proposed ring ancient rarity decrease technique is made out of three stages, viz., (i) antique decrease with the pre-adjustment strategy; (ii) curio decrease with the prepared CNN; and (iii) picture common relationship process with the pre-remedied picture and the CNN yield picture. The ring artifact reduction method consists of three steps. They are pre-correction method that contains artifact reduction mechanism, trained CNN that consists of artifact reduction and the process of correlating mutually the image which contains output of CNN and the pre corrected image. The flow of work is shown in Figure 6.7.

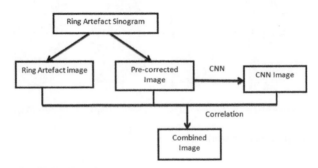

FIGURE 6.7   Workflow of proposed method.

In this concept, a CNN is acquainted into a structure with smother the ring curios. The outcomes have exhibited that, utilizing CNN recognizes curios and improve picture quality. Its constraint is, CNN can scarcely expel all the ancient rarities, and gentle antiques normally remain on the grounds, that it is practiced in the picture space. Looking at the pictures of CNN along with pre-remedy strategy, it is notified that the lingering reasonable ancient rarities don't made available in the pre-remedied picture. The modified perception is clarified by the way that pre-rectification WF technique expels the relics in sinogram space and acquires the recreated picture from adjusted sinogram. In this way, the benefits of sinogram space technique and picture area strategy can be intertwined to create an acceptable joined picture by utilizing picture common connection. Additionally, the presentation of the proposed strategy might be additionally enhanced by bringing the extra pre-adjusted techniques hooked on the system. The effect might be of double folded. To start with, a maximum number of pre-amended strategies gives adequate data to CNN, to recognize antiques and guarantee the presentation of prepared CNN. Second, by utilizing the picture-shared relationship, the benefits of each pre-revised picture can be intertwined to improve the picture quality. The outcomes exhibit that; the suggested method will be performed well for both recreation and genuine exploratory information. It is notified that the CNN for each case in this investigation is equivalent. Later on, preparation can be for the distinctive CNNs for the relating CT pictures, for example, stomach area CT pictures, head CT pictures, small scale creature CT pictures, etc., for improving the presentation that could adjust ring relics in the picture space. Besides modified technique, the information is driven and utilized generally in various CT related applications. To additionally approve the suggested technique's proficiency, all further genuine CT information ought to be utilized. Despite the fact that, proposed technique takes a shot at 2D picture cuts, it very well may be handily stretched out to 3D picture squares. Working on 3D squares might be an effective method to improve the exhibition of the proposed technique. In the interim, preparing with 3D squares will be tedious.

6.4 TUBERCULOSIS (TB) DETECTION WITH CNN WITH LSTM

Tuberculosis or TB, a sickness basically influencing lungs is tainted by bacterium mycobacterium TB and analyzed via cautious assessment of infinitesimal pictures taken from sputum sample. Analysis of malady utilizing microscopy and PC vision strategies are applied for some past functional issues. As of late, profound learning is assuming a significant job in PC vision applications delivering exceptional execution. Be that as it may, computational multifaceted nature consistently stays as a hindrance in the use of profound learning in numerous perspectives. So, a shallow CNN with LSTM (long short term memory) layer is utilized for identifying the tubercle bacillus and mycobacterium TB from the infinitesimal pictures of the specimen gathered from the patients. The predefined model is creating preferred execution over the best-in-class display and furthermore has decreased number of learnable parameters which requires nearly less calculation than the current model [6].

6.4.1   GENERATION OF TEST DATA AND TRAINING DATA

The pathologists who were experts in this field annotate the areas with the hopping boxes in minuscule pictures of sputum test samples. The TB bacilli are observed with the human eye using the software. This dataset might contain 1265 sputum images with hopping boxes around bacilli [7]. Exorbitant nearness of a huge number of negative fixes in the pictures in correlation with the quantity of positive patches makes the model one-sided. So as to take care of this issue, haphazardly chose negative patches were castaway to such an extent that it became multiple times the quantity of positive patches. At that point, new positive patches were consolidated by expanding the first positive patches, for example, by contorting and flipping, making seven extra positive samples for each veritable example. This makes the dataset nearly adjusted, containing 43.7% positive patches. Half of the arbitrarily chosen pictures in the fair dataset are utilized for preparing, and the other half is utilized for testing.

6.4.2   ARCHITECTURE USING CNN AND LSTM

Utilizing a system with completely associated layers isn't usually used to arrange pictures, as a result of the way that such system engineering doesn't think about spatial data of the pictures. It gives a similar significance to the information pixels which are near one another and far separated. This issue presents the CNN with the end goal that, the data about the spatial structure is caught from the preparation information. CNN utilizes uncommon design, which gets adjusted productively for characterizing the sources of info. Plan of CNN incorporates input layer, different shrouded layers, and a yield layer.

Fundamentally, concealed layers comprise of convolutional layer, pooling layer and completely associated layer. The CNN uses the weights shared in the convolutional layer which means the filter that is used earlier is now being used for the local approachable field on the tensor input, that could reduce the amount of parameters used for learning and the performance is improved by using additional layer which is called LSTM (long short-term memory units). The architecture of the LSTM model is shown in Figure 6.8.

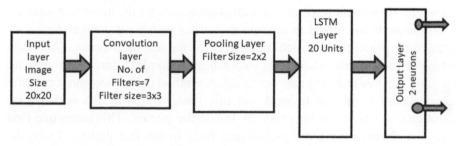

FIGURE 6.8   Architecture for tuberculosis detection.

The image is passed to this convolutional layer, where numbers of channels (pieces) are set as one of the hyperparameters. Pieces characterized are generally arbitrary while preparing without any groundwork, and furthermore, every channel are exceptional. An element map is created by panning the channels over the whole picture. The element map produced by convolution subtleties upgrades portrayals of key highlights in the picture like edges and novel cases. Ensuring the use of other weight channels to picture progressively creates a lot of highlight maps speaking to a similar picture. Pooling is a procedure, which is utilized for the measurement decrease, and it is applied after at least one convolutional layers. Max-pooling activity assists with separating most extreme worth relating to the noticeable highlights present in the element maps. This is likewise done to lessen the component of the element maps.

The LSTM layer is one of the categories of recurrent neural network (RNN) layer, and it is used for pertaining some time-series issues and the sequential information in a model. The only thing that makes this LSTM to be different from that of RNN is the function execution. Some of the LSTM pros are it produces some change in the information by adding and multiplying.

6.4.3   TEST DATA SET DETECTION

After completing the model preparing technique, the created model is sufficiently skilled to name the nearness or nonattendance of pathogen in little picture patches. Above depicted technique is utilized to arrange fixes in a constrained piece of a picture, which comprise objects of intrigue. It is required to isolate the picture into patches, so as to distinguish pathogens in a whole picture or field of view. At that point, these patches are assessed individually utilizing the prepared model, and patches are selected which have driving actuation scores. All things considered; this unassisted strategy alone impacts the model to arrange numerous overlying patches for every unique fix present in the picture utilized for the testing process. This issue emerges particularly when little step length is applied during the making of copy patches bringing near covering. A method called non-most extreme concealment is utilized to work out this issue, which assists with sifting through one enactment for every fix inside the picture. This technique first gets the covers among the picked and fixes in the test picture. From the chosen patches, the arranging method ought to be completed so as to pick the one with the most elevated likelihood score and ousting the others [7].

The created model is tested by using some TB images that contain tubercle bacilli. By examining the test results of image given, it is found that the center row provides high scores which has negative labeled patch providing false detection and this provides annotation errors. Figures 6.9 and 6.10 displays the graph of ROC curve and precision-recall curve respectively, from which it shows true positive rate is more than false-positive rate and Precision-Recall ensembles true positive over true positive plus false positive and true positive over true positive plus false negative.

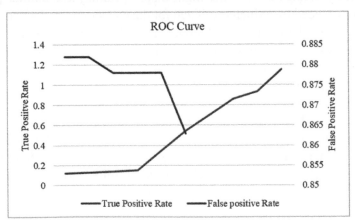

FIGURE 6.9   ROC curve.

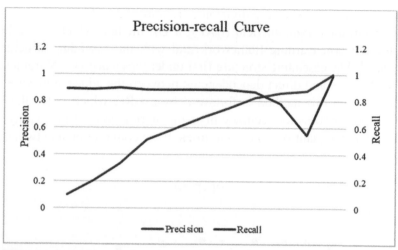

FIGURE 6.10   Precision-recall curve.

LSTMs are ostensibly the best neural system, which is predictable in applying for pictures too. In examination with convolution and pooling layer which just has nearby setting window, LSTM layer builds up the way that each component enactment in the yield is an incitement at the particular point in the entire info 2D tensor. LSTM layer in the system passes data of the whole info grid through parallel associations, while CNN alone adventures just the nearby data. This assists with separating very much pressed component portrayal of the contribution with the assistance of sidelong associations disposing of pointless rehashed highlights at unmistakable areas of the info. This permit the model to tackle slight movements of the highlights over various progressive patches. Primary impediment of the utilization of the LSTM layer is that it obstructs the equal handling due to the consecutive parallel associations. This confinement applies just to demonstrate parallelism and any strategy of information parallelism is feasible for the model. Figure 6.11 sketches the graph of true positive rate versus false-positive rate, in which it depicts TPR, is higher in LSTM models for TB detection.

6.5   CNN-CRF MODEL

Without overlapping, extracting the own feature from an image is an image segmentation process. As a result of fast advancement in the field of clinical imaging, exactness in picture division is required. For this purpose, for the use of conditional random field (CRF), a DL method which is combined with the uniform learning framework and with the cascading structure is established. The cascading structure first undergoes a deep CNN framework for the effective and direct dependencies between the adjacent closure tags available. Then this method is followed by a CRF, which is for the later segmentation process for accuracy in segmentation, system profundity, and to recognize the quantity of pooling times in the convolutional network [8].

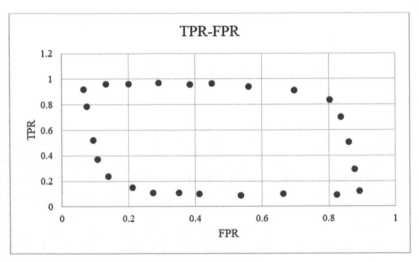

FIGURE 6.11   TPR-FPR curve for tuberculosis detection.

Segmentation algorithm of this method makes use of subnetwork structure in the neural network. Multiple convolutional networks efficiency needs to improved, which is possible by directly simulating the adjacent tags dependence of the segmentation. This process results in the prediction of energy which is influenced by the model value of the tags that are presented adjacent to each other. To produce the output DCNN (Deep CNN) make use of cascade structure which is implemented using two convolutional networks in which initial convolutional network acts as the input for the next convolutional network. A residual connection is used for the enabling the DCNN for training more effectively and efficiently. After training, linking with an independent parameter layer will be created in the network.

The resultant chart which is created by the CNN for relapse is moderately questionable. With the assistance of restrictive likelihood displaying the inadequacies which exist in CRF can be improved. CRF is a contingent irregular field, which incorporates a restrictive likelihood appropriation model that acknowledges an info arbitrary factors for which a yield of arbitrary variable is delivered. CRF is undirected discriminative likelihood diagram model and contingent likelihood dispersion. This CRF technique is, for the most part generally utilized in fields like normal language preparing and picture handling. CNN-CRF division technique structure chart is revealed in Figure 6.12.

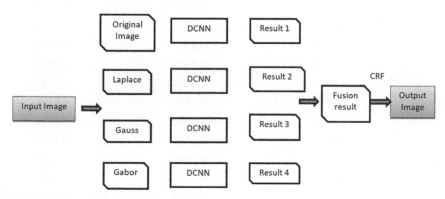

FIGURE 6.12   CNN-CRF structure.

For brain tumors diagnosis, a novel division technique for MRI division is utilized. This division procedure includes the element extraction and convolutional neural system together for the finding of cerebrum tumors. The last division of this MRI picture classified dependent on the size, shape, and the situation of brain tumors. The pictures are separated first with various modular highlights utilizing CNN, then the whole sectioned images are combined together for the resultant image. By using GMM (Gaussian Markov model), the fused resultant image is post-processed for the final segmentation image. When compared to the conventional method, this method is more accurate and makes use of the basic convolutional neural network (CNN). The main drawback of this CNN method is its network structure which cannot pretend the undeviating requirement that is found amongst the spatial closure tags. So, the CNN segmentation is modified by using DCNN along with the cascade structure which includes three remaining layers and ASPP (atrous spatial pyramid pooling) model. For better comprehending the inconsistency between division precision and system profundity, the pooling technique is utilized. CRF is utilized for the post handling division in the deep convolutional network. Table 6.5 lists out the segmentation results with parameters dice coefficient, positive predictive value, and sensitivity followed by Figure 6.13 shows the graphical representation of segmentation results.

TABLE 6.5 Segmentation Results

Number	Dice Coefficient	Positive Predictive Value	Sensitivity
1.	0.9288	0.8697	0.9136
2.	0.7858	0.9321	0.9077
3.	0.8398	0.8062	0.8752
4.	0.8685	0.9156	0.7963
5.	0.8237	0.9269	0.8950
6.	0.8998	0.8364	0.8995
7.	0.9353	0.9007	0.8837
8.	0.8934	0.9501	0.8558
9.	0.8266	0.9642	0.8673
10.	0.8085	0.8841	0.8981
11.	0.8454	0.9369	0.8527
12.	0.8527	0.8725	0.9053
13.	0.9135	0.8943	0.8935

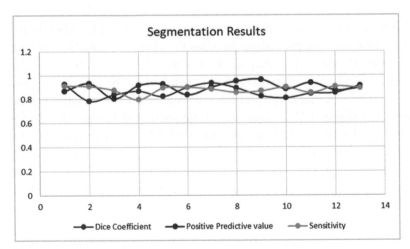

FIGURE 6.13   Graph of segmentation results.

The MRI image capture for the brain tumor is a 3D representation, because of less resolution in the three dimensions, the image is clipped in the direction along the feature axis, which results in the image with a two-dimension slice. At the image block level, the image convolution takes place and 33×33 two-dimensional data blocks are obtained for the purpose of training the data. When the brain area is segmented manually then the area contains five parts. They are no tumor, edema, corruption, non-supported development of the tumor, and nonstop development of the tumor. In these five localities, just three is considered with the end goal of execution assessment. By consolidating the element combination and DCNN mind tumor MRI division strategy, the unpredictability in the single system will be diminished and the issue in the process like classification of accuracy and segmentation can be avoided.

6.6 CAD-CNN

Efficient diagnosis and treatment process is the main aim of medical image analysis. There are many methods through which this is possible, and one such method is the computer-aided detection or diagnosis (CAD). The rapid development in the available technological field and the manual data interpretation in CAD make it a challenging task. To solve this problem, a method in DL, i.e., CNN is successfully utilized. For detecting the prostate cancer, breast cancer diagnosis and for lung nodule detection, this CNN is effectively casted-off. There are four categories of these DL methods viz., CNN-based methods, sparse coding, autoencoders (AEs), and restricted Boltzmann machines (RBMs). Among this the method which has more and more attraction in and around the world by its achievable and promising results is CNN-based methods [9].

There are two major parts of the computer-aided detection or diagnosis. Lesion detection and false-positive reduction are two different parts. For the detection task, based on specific algorithms lesion part is utilized. This lesion part results in lesions of applicant. For reducing the false positive rate, the second part viz., false-positive reduction part makes use of traditional machine learning algorithms. The complex and sophisticated programs does not give a better performance in the conventional CAD, which results in less usage of this method. The DL technique, which is widely known as CNN base method, gives a solution for this problem in CAD. The distinguished character of this method provides various systems for the detection of various diseases and different image modalities. Figure 6.14 displays the workflow of the CAD system.

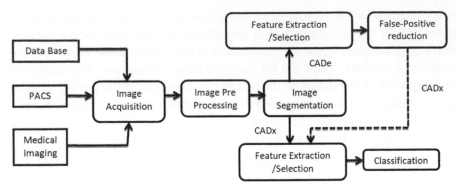

FIGURE 6.14   Workflow of a typical CAD system.

The very first application is breast cancer detection and diagnosis by numerous medical imaging modalities and CNN based methods. The test has been carried out for around 171 test set for tumors, the sensitivities achieved are 95%, 90%, 85% and 80% at 14.03, 6.92, 4.91 and 3.62 false positives per patient (with six passes), respectively. Based on the result, it is clear that this method is more feasible since the number of false-positive at 100% sensitivity made to less and reduced further.

The next application is for lung cancer detection. The most frequent and the disease which leads to death all over the world is this lung cancer. In order to improve the survival rate, the only way is the efficient lung cancer detection at the beginning stage as soon as possible in the form of lung nodules. The lung CAD system is divided into two categories one for lung nodules detection and another for classification. The lung image DB association and lung image catalog associates the image collection DB used for the validation and verification of results.

Prostate cancer is the next application of the CNN based method. Prostate cancer is considered as the most common disease that could occur among the men and which is the leading deadly disease among men. A method named region-based CNN framework for detection of prostate cancer is used. This method provides an accuracy of 99.07% and an average AUC of 0.998 with performance in epithelial cells detections.

Fields like lesion segmentation, detection, and classification of medical image analysis make use of CNN-based methods. This method is effectively used in spite of all the limitations like data augmentation and learning. This method is greatly benefited in the medical image analysis in the detection of various diseases and also in radiomics, precision medicine, and imaging grouping.

6.7 CNN-BASED ORIENTATION CLASSIFIER

For the evaluation of stenosis and atherosclerotic plaque, the cardiac CT angiography is going to be the prerequisite. CNN uses an algorithm for the extraction of coronary artery centerlines in cardiac computed tomography angiography (CCTA). The CCTA picture dependent on a nearby picture fixes and for anticipating the most probable bearing a span of supply route at some random point 3D expanded CNN is equipped. The procedure begins at a solitary seed point set either physically or consequently in and around coronary supply route, trackers follows the vessel centerline in two ways utilizing the forecast of the CNN. The procedure will end just when there isn't a course is related to high sureness. The CNN is physically prepared utilizing centerlines in formulating pictures. The whole procedure is exclusively guided by the nearby picture thus no picture preprocessing isn't needed [10].

The CNN model was at first organized by a preparation set containing 8 CCTA pictures with physically clarified centerlines and a sum of 32 offered in coronary conduit following test (CAT08). The assessment procedure was performed utilizing a test set comprising of 24 CCTA test pictures in which 96 centerlines were removed. The centerlines removed will have a cover with a normal of 93.7% with manual reference centerlines. The focuses that are extricated will have profoundly exact and with a normal separation of 0.21 mm of the reference. In view of assessment, this technique will be most likely utilized. In this, the picture is viewed as whose direction and sweep of the coronary corridor are situated at the area x, which depends on 3D isotropic picture and single CNN. The assessed yield of the CNN layers contains numerous grouping of modes that are utilized for the assurance of likelihood appropriation over the course D and utilized for the assurance of span r of the seed point x.

For CAT08 challenge the CNN is prepared with 3D CCTA pictures. The reference pictures after assessment gives four centerlines. The reference explanations incorporate arrangement of focuses, which are utilized for comparing span estimations. Preparing a picture would be a tedious and dreary procedure since all the parts of the coronary vein are physically executed and are comprehensive physically.

To prepare the CNN tests, a reference mark with an adequate conviction is required; at that point, the preparation focuses in those territories where coronary centerlines have been explained. For the estimation of direction and span estimation the CNN engineering is practically indistinguishable. The yield layer comprises just relapse in the last layer as opposed to consolidating characterization and relapse. The inside voxel of the info coronary supply route is anticipated by the estimation of CNN. The blunder of relapse misfortune that exist between the reference and anticipated nearness esteems is limited by this procedure. After this, reference esteem is gotten as the conclusive outcome.

By utilizing the CAT08 challenge information, the centerline extraction was preceded as the initial step of quantitative examination. The reference centerline permits an assessment of centerline cover and precision. During the preparation procedure, eight informational indexes are prepared. For preparing every one of the accessible eight pictures, a CNN was prepared utilizing the other seven pictures. The centerline extraction process was instated at a point which is exceptionally employed for the recognizable proof of coronary conduit.

So as to decide the direction and size of the coronary course, it doesn't requires some other extra highlights, yet rather than it, this technique utilizes CNN to separate the required data straightforwardly from CCTA pictures. Another technique through which some required data can be gotten is via preparing the CNN by exploiting scantily explained vessel trees. The data identified with bearing of the coronary centerline and span of the coronary lumen can likewise be gotten by the CNN using a solitary vessel seed-based iterative tracker at some random point.

This profound learning-based technique is well appropriate for the coronary vein centerline extraction. The comprehension is made such that any convolutional neural system can be made to and at the same time gain proficiency with procedure for assurance of the bearing identified with coronary vein centerlines and sweep of the coronary lumen. This data can be evaluated by employing this CNN with high precision and with fast.

6.8   CNN-BASED SURVIVAL MODEL FOR PANCREATIC DUCTAL ADENOCARCINOMA

For the survival analysis, the most commonly used method is the CPH model. In the field of feature reduction and in modeling, the CPH plays a vital role. This model will limit the prognostic performance by using linear assumptions. The CT images of pancreatic ductal adenocarcinoma patients were captured and tested with the model build by a CNN based survival model. In PDAC prognosis the CPH based radiomics pipelining is performed on the output signal by this CNN based survival model. By overcoming the limitations of conventional survival model, this CNN based model provides better survival patterns on the CT images.

A six-layered convolutions architecture of CNN is considered as the architecture of CNN survival model. The image having dimensions 140×140×1 (in grayscale) is considered as the input image. This input CT image has the manual contours of the tumors. It is assumed that the entire image outside the region is 0 and inside the region varies from 0 to 255 grayscale values. There is a kernel with 32 filters of size 3×3 in all the convolutional layers which is followed by BN (normalized layers). The next component will be the maximum pool layer, in which the first maximum pool layer size is 2×2 and the second one have the pool size as 3×3. The trainable parameters of the model are reduced because of this pool layer. A drop out layer is connected with the small sampled size which is used to avoid overfitting. Figure 6.15 displays the CNN survival architecture blocks.

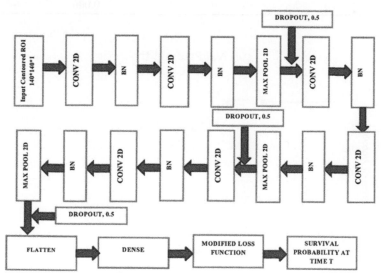

FIGURE 6.15 The projected CNN-survival architecture.

The previously mentioned propelled highlights are contributed towards the expansion in the presentation of the model. Looking at the exchange learning highlights of CPH model and CNN-Survival mode, the CNN endurance mode has higher IPA [19]. From this outcome, unmistakably the misfortune work in this CNN endurance model plays out the customary straight CPH which is generally utilized. Table 6.6 lists out the CNN-Survival probability of a patient deceased 316 days after medical procedure, followed by its graphical representation in Figure 6.16.

TABLE 6.6 Endurance Likelihood Bend Created by the Proposed CNN-Survival Model for a Patient Deceased 316 Days after Surgery

Days for CT	Survival Probability
0	1.00
50	1.00
100	1.00
150	1.00
200	0.75
250	0.5
300	0.01
350	0.008
400	0.006

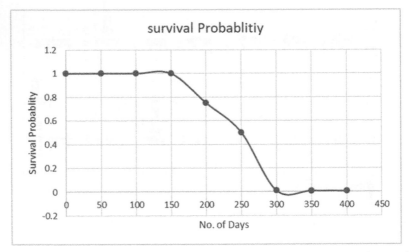

FIGURE 6.16 Graphical representation of CNN-survival model for patient deceased 316 days after surgery.

Through the CT images of three independent cohorts, CNN survival model is being validated in terms of its loss functions. The CNN survival model is meant to avoid the drawbacks of conventional radiomics based CPH model and also it's outperformance with relatively small sample size settings. Validation of the loss function for other types of diseases is also possible by transfer learning process. In the field of quantitative medical imaging, the CNN survival model has huge potential to acts as a standardized survival model. The survival probability of the patient more than one year after the surgery is listed out in Table 6.7 and its graphical form in Figure 6.17.

TABLE 6.7 Endurance Likelihood Bend Created by the Proposed CNN-Survival for a Patient Endure Over One Year after Surgery

Days for CT	Survival Probability
0	1.00
50	1.00
100	1.00
150	1.00
200	1.00
250	0.95
300	0.90
350	0.85
400	0.80

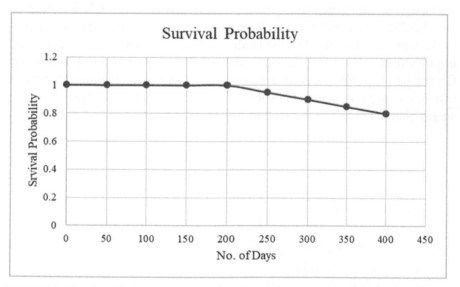

FIGURE 6.17 Graphical representation of CNN-survival model for patient survived more than a year after surgery.

The PDAC prognosis is formed as the result of output produced from CNN based survival model, which is a traditional CPH based radiomics and transfer learning method. This provides a better solution for the patterns based CT image survival and acts as a solution for all the limitations that are found in conventional survival models.

6.9   CONCLUSION

In this study, various models prescribed for reducing the computational complexity is analyzed. Here a five-layer CNN, that has six completely connecting learning model is casted-off for the classification of MRI images to identify AD. This is achieved by taking the T2 data form the whole data set for the analysis. The results obtained indicate that the combination of MCI with CN will increase the accuracy up to 98%. Another CNN based approach is discussed that overcome some of the limitations observed from analyzing the CT images. The algorithm that uses trained CNN for reducing the ring artifacts are studied, and it helps to reduce artifacts by providing a hybrid structure that combines CNN along with pre-corrected images. Thus the image quality is improved as this hybrid nature of the algorithm reduces the artifacts. Then for determination of some crucial diseases, CNN combined with LSTM is used and it shows the best performances ever when compared to the previous models. Thus the analysis of various models representing an improvement in the computational accessibility is discussed in this chapter. On the whole, CNN provides a superior platform with its inimitable algorithms for medical image classification and disease identification in earlier stages, thereby reducing the fatal rate of mankind.

KEYWORDS

- **Alzheimer's disease**
- **artifacts**
- **cardiac computed tomography angiography**
- **conditional random field**
- **deep learning**
- **Gaussian Markov model**

REFERENCES

1. Basheera, S., & Satya, S. R. M., (2020). A novel CNN based Alzheimer's disease classification using hybrid enhanced ICA segmented gray matter of MRI. *Computerized Medical Imaging and Graphics*, 101713.

2. Goceri, E., & Caner, S., (2017). Automated detection and extraction of skull from MR head images: Preliminary results. In: *2017 International Conference on Computer Science and Engineering (UBMK)*. IEEE.

3. Evgin, G., (2018). Comparison of weighted K-means clustering approaches. *International Conf. on Mathematics (ICOMATH2018)*.

4. Chang, S., et al., (2020). A CNN based hybrid ring artifact reduction algorithm for CT images. *IEEE Transactions on Radiation and Plasma Medical Sciences*.

5. Münch, B., et al., (2009). Stripe and ring artifact removal with combined wavelet: Fourier filtering. *Optics Express, 17*(10), 8567–8591.

6. Simon, A., et al. (2017). *Shallow CNN with LSTM Layer for Tuberculosis Detection in Microscopic Image*.

7. Quinn, J. A., et al., (2016). Deep convolutional neural networks for microscopy-based point of care diagnostics. *Machine Learning for Healthcare Conference*.

8. Feng, N., Xiuqin, G., & Lijuan, Q., (2020). Study on MRI medical image segmentation technology based on CNN-CRF model. *IEEE Access*.

9. Gao, J., et al., (2019). Convolutional neural networks for computer-aided detection or diagnosis in medical image analysis: An overview. *Mathematical Biosciences and Engineering, 16*(6), 6536.

10. Wolterink, J. M., et al., (2019). Coronary artery centerline extraction in cardiac CT angiography using a CNN-based orientation classifier. *Medical Image Analysis, 51*, 46–60.

11. Khalvati, F., et al., (2019). *Radiomics*, 597–603.

12. Van, G., & Joost, J. M., et al., (2017). Computational radiomics system to decode the radiographic phenotype. *Cancer Research, 77*(21), e104–e107.

13. Khalvati, F., et al., (2019). Prognostic value of CT radiomic features in resectable pancreatic ductal adenocarcinoma. *Scientific Reports, 9*(1), 1–9.

14. George, B., Samantha, S., & Inmaculada, A., (2014). Survival analysis and regression models. *Journal of Nuclear Cardiology, 21*(4), 686–694.

15. Keek, S. A., et al., (2018). A review on radiomics and the future of theranostics for patient selection in precision medicine. *The British Journal of Radiology, 91*(1091), doi: 10.1259/bjr.20170926.

16. Klekociuk, S. Z., et al., (2014). Reducing false-positive diagnoses in mild cognitive impairment: The importance of comprehensive neuropsychological assessment. *European Journal of Neurology, 21*(10), 1330-e83.

17. Weissberger, G. H., et al., (2017). Diagnostic accuracy of memory measures in Alzheimer's dementia and mild cognitive impairment: A systematic review and meta-analysis. *Neuropsychology Review, 27*(4), 354–388.

18. LeCun, Y., Yoshua, B., & Geoffrey, H., (2015). Deep learning. *Nature, 521*(7553), 436–444.

19. Zhang, Y., et al., (2020). CNN-based survival model for pancreatic ductal adenocarcinoma in medical imaging. *BMC Medical Imaging, 20*(1), 1–8.

CHAPTER 7

EXPLORATION OF DEEP RNN ARCHITECTURES: LSTM AND GRU IN MEDICAL DIAGNOSTICS OF CARDIOVASCULAR AND NEURO DISEASES

R. RAJMOHAN,[1] M. PAVITHRA,[2] T. ANANTH KUMAR,[2] and P. MANJUBALA[1]

[1]Associate Professor, Department of CSE, IFET College of Engineering, Gengarampalayam, Tamil Nadu, India, E-mails: rjmohan89@gmail.com (R. Rajmohan), pkmanju26@gmail.com (P. Manjubala)

[2]Assistant Professor, Department of CSE, IFET College of Engineering, Gengarampalayam, Tamil Nadu, India, E-mails: pavimuthu27@gmail.com (M. Pavithra), ananth.eec@gmail.com (T. A. Kumar)

ABSTRACT

Deep learning (DL) algorithms utilize an assortment of factual, probabilistic, and advancement strategies to gain from past understanding and recognize helpful examples from enormous, unstructured, and complex clinical datasets. The extent of this exploration is fundamentally on the investigation of infection expectation approaches utilizing various variations of discriminative profound learning calculations. In this perspective, the chapter intends to deliver specific indispensable material about RNN-based DL and its solicitations in the pitch of biomedical engineering. This chapter inspires young scientists and experts pioneering in the biomedical domain to swiftly comprehend the best-performing methods. It also empowers them to associate diverse RNN approaches and move towards the future of healthcare provisioning in human life.

7.1 INTRODUCTION

Artificial intelligence (AI) is the simulation of the intelligence of the human thoughts that can be applied to any machine to show and perform the attributes of humans, such as problem-solving and learning the information. In mimicking the human brain, deep learning (DL) procedures are used as a subset of the simulated intellect. Nowadays, DL techniques play a major role in predicting the diseases. DL calculations have a wide scope of uses, including computerized content arrangement; organize interruption recognition, garbage e-mail sifting, identification of Visa misrepresentation, client buy conduct discovery, advancing assembling procedure, and infection demonstrating [1]. In recent times DL approaches have led to a resurgence of neural link-based models. Founding revisions presented stacked restricted Boltzmann machines (RBMs) and stacked auto-encoders, which exhibited striking enactment in image processing, retaining the layer-wise pre-training procedure. Since then, variants of neural network solicitation have reconnoitered deep architectures in the medical field, industry automation, and among other fields.

DL is classified by two methods, discriminative learning (DL) and generative learning (GL). Recurrent neural network (RNN) is one of the methods in the DL used to convert the data in the order of the chain series. RNN faces the vanishing and exploding problems. To overcome this issue, LSTM and GRU models are proposed. The majority of these applications have been executed utilizing discriminative variations of the profound learning calculations as opposed to generative ones. In the discriminative variation, a forecast model is created by learning a dataset where the mark is known, and appropriately, the result of named models can be anticipated.

Illness forecast and in a more extensive setting, clinical informatics, have as of late increased critical consideration from the information science examine network as of late. This is fundamentally because of the wide adjustment of PC-based innovation into the wellbeing division in various structures (e.g., electronic wellbeing records and authoritative information) and ensuing accessibility of huge wellbeing databases (DBs) for specialists. This electronic information is being used in a wide scope of human services look into zones, for example, the examination of social insurance usage, estimating execution of clinic care arrange, investigating examples and cost of care, creating sickness chance forecast model, ceaseless malady observation, and contrasting illness pervasiveness and medication results. Our examination centers around the malady chance forecast models including profound learning calculations (e.g., LSTM, GRU, Stacked LSTM, Bidirectional LSTM, BiStacked LSTM, Stacked GRU, Bidirectional GRU, BiStacked GRU), explicitly-discriminative learning calculations. Models dependent on these calculations utilize named preparing information of patients for preparing [2].

Given the developing pertinence and adequacy of discriminative AI calculations on prescient ailment displaying, the expansiveness of research despite everything appears advancing. This exploration intends to distinguish key patterns among various kinds of discriminative AI calculations, their exhibition exactness, and the sorts of ailments being examined. Likewise, the points of interest and confinements of various administered AI calculations are abridged. The consequences of this examination will assist the researchers with bettering comprehend momentum patterns and hotspots of malady forecast models utilizing discriminative AI calculations and define their exploration objectives likewise.

Among various managed AI calculations, this investigation looked into, by following the PRISMA rules [3], existing examinations from the writing that utilized such calculations for malady expectation. The primary commitment of this examination (i.e., the correlation among various administered profound learning calculations) is increasingly exact and far-reaching since the correlation of the presentation of two diverse calculations across various investigation settings can be one-sided and create incorrect outcomes. Generally, standard factual strategies and specialist's instinct, information, and experience had been utilized for anticipation and ailment chance expectation. With the expanding accessibility of electronic wellbeing information, progressively powerful and progressed computational methodologies, for example, AI has gotten increasingly down to earth to apply and investigate in the infection forecast region. In the writing, the vast majority of the related investigations used at least one AI calculations for a specific illness expectation. Consequently, the exhibition examination of various profound learning calculations for infection expectation is the essential focal point of this investigation.

Coronavirus, otherwise called COVID-19 is a pandemic sickness in the present circumstance which starts toward the finish of 2019 in China. The flare-up of the infection is worldwide, and the influenced cases are around 3,000,000 announced by the World Health Organization (WHO). Appropriate inoculation for COVID-19 has been still looked into by numerous analysts around the world. The main analysis strategy accessible for COVID-19 is RTPCR (reverse transcription-polymerase chain reaction), Kit. Be that as it may, this unit doesn't bolster for the early recognition of maladies and treatment [21]. Furthermore, COVID-19 is said to an asymptotic illness, the nature of coronavirus is it demonstrates the side effects to the individual following 13–17 days it goes into the person in question. Be that as it may, 80% of cases are asymptotic cases revealed around the world, it doesn't cause any manifestations, yet the test outcome is certain. In this way, it is hard to recognize COVID-19 sickness. As of late, numerous inquiries have been accounted for by the analysts worldwide to recognize the COVID-19 at the beginning period utilizing the headway of the AI. The main thing required by AI calculations is information. In view of the information prepared to the neural system, it gives the test outcome all the more precisely. RNN is one of the profound learning calculations which prepare the systems dependent on the pictures. Inside the brief time frame, numerous works are accounted for dependent on RNN and COVID-19.

RNN can be utilized to predict the futuristic inferences of coronavirus. The effect of COVID-19 is worldwide and step-by-step the passing rate and contaminated pace of the infection has been expanded quickly. What's more, still there is no immunization, and strategies to identify the ailment at a beginning period have not been demonstrated despite the fact that numerous sorts of research have been accounted for. The sickness can't be distinguished at beginning periods in light of the absence of indications, and even numerous asymptotic patients are accounted for around 80% in India. Scientists report numerous profound learning and AI-driven calculations to recognize the infection at a beginning period; however, there is just a base measure of datasets that will prompt better Precision. It is simple for profound learning calculations to prepare few datasets and anticipate the testing exactness. Inside four months, the infection spread worldwide and influences the endurance of people and creatures. Specialists are doing numerous sorts of examinations to control the spreading and recognizing the immunization for COVID-19. In the innovative side, analysts attempt to foresee the sicknesses in beginning times utilizing a few procedures or calculations.

7.2   RECURRENT NEURAL NETWORK

RNN is a model of neural schema that profits the worth yield from the neural system back to the information. Along these lines, the yield estimation of the output is transferred rear to the neural system, with the goal that the weight computation of each time purpose of the system model is identified with the substance of the past time point, which implies that the neural system is incorporated. The idea of time, through such an instrument, makes the neural system memory. Figure 7.1 depicts the repetitive neural system engineering of RNN. After each round of activity, the yield of the web is held up to the following exchange activity. In this way, the time i will be $i+1$ time focuses. The yield is contemplated, so it has the attributes of when memory. The following equation depicts the mathematical model of RNN [4]:

$$a_i = \partial_a(Z_a m_p + V_a n_{i-1} + c_j) \tag{1}$$

$$n_i = \partial_n(Z_n a_i + c_n) \tag{2}$$

where; m_p is the input layer vector, a_i is the hidden layer vector, n_i is the output layer vector, Z, V, and care weight parameter matrix, and Z_a and Z_n are the activation functions.

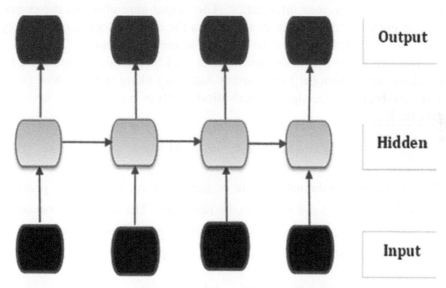

FIGURE 7.1 RNN neural engineering system.

RNN models are normally fit to worldly sequenced information, and a few variations have been produced for sequenced highlights. Hochreiter and Schmidhuber [5] proposed a long momentary memory called LSTM, displaying great execution in various grouping-based errands, for example, penmanship acknowledgment, acoustic demonstrating of discourse, language demonstrating, and language interpretation. Cho et al. [6] recommended the gated repetitive unit called GRU model, fundamentally like yet easier than LSTM, and demonstrated similar, if worse, execution.

7.2.1   LONG SHORT TERM MEMORY

In the plinth of DL, LSTM is one of the techniques of RNN. The gates such as the input gate, output gate, forgot gate and memory cell makes the architecture of LSTM. It is sketched to outfit the long-term dependencies problem, and by using the separate memory cell to store the data, the vanishing problems are avoided. It uses cell states to carry the information in the long term even it is not used. LSTM uses three types of gates, namely forget, update, and output to access the data [5].

The first gate called forgot gate, is used to remove the unwanted data from the previously stored data. The update gate is used to insert the required data along with the previous information. The output gate provides the current data or output to the next layer to process the data for long-term memory. The three gates help to move the data in a sequential manner.

The forgot gate (fg_l) is the first layer in the LSTM which concatenate the previous information (pi_d) and current data (cd_d) along with the load (w(f)) and bias (b(f)) and the sigmoid activation function is used to run the forgot gate [6, 32].

$$fg_l = \sigma(wi(fg)\,(pi_{d-1}, cd_d) + bi(fg)) \qquad (3)$$

The second gate is the update gate or input gate (ui_d), where the previous information is added and passed to the next layer. The input gate consists of the separate load (w(i)) and bias (b(i)) which get combined with previous and current data along with the following function.

$$ui_d = \sigma(wi(ui)(pi_{d-1},\, cd_d) + bi(ui)) \qquad (4)$$

Next the new candidate ($c_d^\sim$) value is added to the input gate with the tanh function (Figure 7.2).

$$c_d^\sim = \tanh(wi(nc)(pi_{d-1},\, cd_d) + bi(nc)) \qquad (5)$$

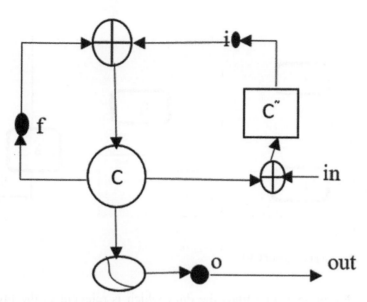

FIGURE 7.2   Architecture of LSTM.

The third step is to update the data with forgetting unwanted data and by adding new data.

$$c_d = fg_1 * cd_{d-1} + ui_d * nc_d\tilde{}$$ (6)

The output layer that is the final layer of the LSTM architecture decides what kind of information will be fed as input to the next layer. The output is multiplied with the updated data.

$$ol_d = \sigma(wi(ol) [pi_{d-1}, cd_d] + bi(ol))$$ (7)

$$hi_d = ol_d * (\tanh(c_d))$$ (8)

7.2.2   GATED RECURRENT UNIT (GRU)

GRU [6] is similar to the LSTM that was designed to capture the dependencies of more time scaling with the help of the recurrent unit in the architecture. As LSTM, GRU consists of gating systems that fine-tune the flow of data inside the cell which does not have any segregate memory cells (Figure 7.3).

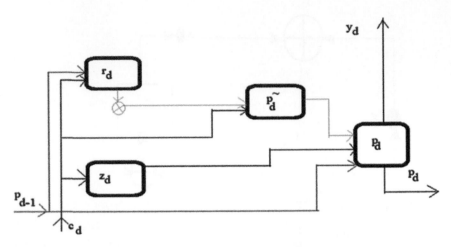

FIGURE 7.3   Architecture of GRU.

The relevant gate (rg_d) finds the data which is relevant to the previous data and the update gate (zg_d) determines the removal and updating of the data [6]:

$$rg_d = \sigma(wi(rg)\,[pi_{d-1},\, cd_d]) \qquad (9)$$

$$zg_d = \sigma(wi(zg)\,[pi_{d-1},\, cd_d]) \qquad (10)$$

The temporary value ($p_d^{\sim}$) is created by multiplying with relevant data with the tanh activation function:

$$p_d^{\sim} = \tanh\,(wi.rg_d{}^*\,[pi_{d-1},\, cd_d]) \qquad (11)$$

The last gate (PD) is the output gate from which the necessary data is transformed into the next layer:

$$p_d = (1-d)ZD * p_{d-1} + z_d * p_d^{\sim} \qquad (12)$$

The advantage of the gated recurrent unit (GRU) is the reduced number of parameters as compared with the long short term memory (LSTM). The survey to predict heart disease is taken between these two techniques of DL.

7.3   ACTIVATION FUNCTIONS FOR RNN

The normalization process of the output is carried over by the activation functions. In a neural network, every layer is composed of a set of neurons. The purpose of every neuron is to execute a non-sequential transformation on the input to produce output (Figure 7.4).

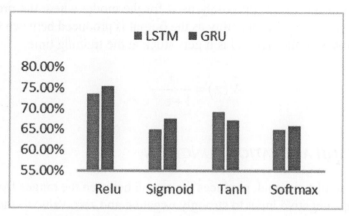

FIGURE 7.4   Comparison of activation function.

The activation function, such as sigmoid, tanh, relu, and softmax are compared between LSTM and GRU [7]. Out of these functions ReLU gives the best result in the activation functions. The survey Table 7.1 is given.

TABLE 7.1   Determination of Activation Function

Activation Function	LSTM	GRU
ReLU	73.62%	75.59%
Sigmoid	64.96%	67.72%
Tanh	69.29%	67.32%
SoftMax	64.96%	65.96%

7.3.1   RECTIFIED LINEAR UNIT (RELU)

Relu allows the non-linear activation function, consist of a derivative function that allows backpropagation, which weight and bias can change the accuracy of the output results.

$$A'(z) = \begin{cases} 0 \ for \ z < 0 \\ z \ for \ z \geq 0 \end{cases} \tag{13}$$

7.3.2 SIGMOID

The sigmoid function is purposely used for the model where the prediction of output should be in probability as the output is produced between 0 and 1. The drawback of the sigmoid is it gets stuck at the training time.

$$A'(z) = \frac{1}{1+e^{-z}} \tag{14}$$

7.3.3 TANH ACTIVATION FUNCTION

Tanh is also like sigmoid, produces the output between the ranges from −1 to +1. It maps negative input to strongly negative and zero value to the origin. The main thing is it does not allow for backpropagation.

$$\tanh(z) = \frac{2}{1+e^{-2z}} - 1 \tag{15}$$

7.3.4 SOFTMAX ACTIVATION FUNCTION

SoftMax function can handle the multiple classes. The advantage of the sigmoid activation function is, it does not operate all the neurons and make the network plain making it well organized and easy for estimation. Therefore, the sigmoid function is used for further surveys such as loss function, optimizer, and hidden layer.

7.4 LOSS FUNCTIONS FOR RNN ARCHITECTURES

In the artificial neural network (ANN), loss function plays an important role, which is utilized to measure the conflicting between the predicted data and original data. Loss function [8] maps the set of parameters value for the network onto a scalar value that indicates how well those parameters accomplish the tasks. The survey between the different loss functions is taken between the LSTM and GRU. Mean_squared_error gives better accuracy among the other loss functions (Table 7.2 and Figure 7.5).

TABLE 7.2 Determination of Loss Function

Loss Functions	LSTM	GRU
Mean_squared_error	64.93%	66.96%
Squared_hinge	34.04%	35.04%
Cosine Proximity	61.90%	63.96%

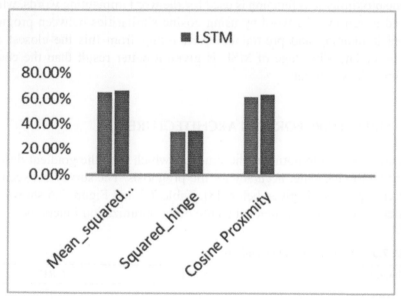

FIGURE 7.5 Comparison of loss function.

7.4.1 MEAN_SQUARED_ERROR (MSE)

Mean_squared_error is one of the loss functions is always sensitive towards the outliers. MSE is always non-negative values closer to zero or better value.

$$MSE\left(Z,Z\right) = \frac{1}{n}\sum_{j=0}^{M}\left(z_{j} - z_{j}^{\wedge}\right)^{2}$$

$$(16)$$

7.4.2 SQUARED_HINGE

Squared_hinge loss function is mainly used in a binary classification problem. By this loss, the decision is taken as yes/no. It does not show how the classifier is used for the classification.

$$SH\left(Z, Z^{\wedge}\right) = \sum_{k=0}^{n}\left(\max\left(0, 1 - z_j . z_j^{\wedge}\right)^2\right) \tag{17}$$

7.4.3 COSINE PROXIMITY

Cosine proximity loss function is used for the word implanting words, which is used to retrieve the word by using cosine similarities between prognosticated implanting and pre-trained word fixing from this the closest one is chosen. The advantage of MSE is given a better result than the cosine proximity loss function.

7.5 OPTIMIZERS FOR RNN ARCHITECTURES

Optimizer is used to optimize the function which uses the gradient descent algorithm to give better accuracy for the prognosis. The survey between the different optimizer is given below [8]. Table 7.3 and Figure 7.6 shows that Adadelta gives better accuracy than the other optimization functions.

TABLE 7.3 Determination of Optimizers

Optimizer	LSTM	GRU
Adam	71.26%	74.02%
Adagrad	72.44%	72.83%
Adadelta	72.44%	76.36%

7.5.1 ADAM OPTIMIZER

Adaptive moment estimation (Adam) adapts the learning rate (LR) and updates the weights of the network continually based on the training set of data. It implements the epidemic moving average of the gradients to scale the LR. The con of Adam is it has an exponentially decaying average of the past values.

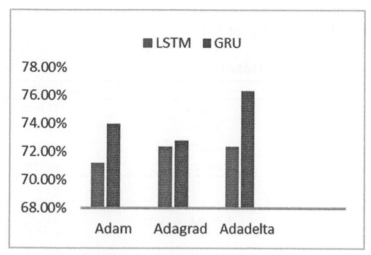

FIGURE 7.6   Comparison of optimization function.

7.5.2   ADAGRAD OPTIMIZER

Adaptive gradient algorithm (Adagrad) adapts the LR-based on the param-eters. Adagrad improve the work on the spare gradient problem. For each and every parameter it uses different LRs-based on previous gradients. The LR always decreases and decayed-this is the main downside of the Adagrad.

7.5.3   ADADELTA OPTIMIZATION

Adaptive LR method (Adadelta) dynamically adapts the LR and reduces the aggressive of the Adagrad optimization. Adadelta reduces the LR by confining the window of the previously collected gradient to a fixed window (w). It does not take the default-LR as it calculates the running average of past data and the current data.

7.5.4   HIDDEN LAYER

Hidden layer is the transitional layer between the input layer and the output layer [9]. It consists of the set of neurons where the activation functions provide the output along with the weights of the input layer. By providing different numbers of hidden layer, the comparison is taken between LSTM and GRU (Table 7.4 and Figure 7.7).

TABLE 7.4   Determination of Hidden Layers

Hidden Layer	LSTM	GRU
5	64.96%	71.96%
3	65.8%	70.96%
3	66.34%	68.98%
2	68.11%	66.15%

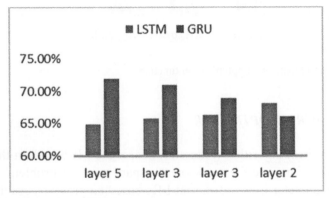

FIGURE 7.7   Comparison of hidden layer.

The more number of hidden layers will give more accuracy in the Recurrent Neural Network (RNN) techniques.

7.6   VARIANTS OF RNN MODELS

7.6.1   STACKED LSTM

The very first defined LSTM model which followed the protocol and standard defined by the feed network is the stacked LSTM structure [10]. The overall structure of a stacked LSTM is comprised of numerous learning layers which are made of memory units. Stacking LSTM concealed layers makes the model further, more precisely winning the depiction as a profound learning strategy. Each layer frames some bit of the task to understand, and offers it to the accompanying. In this sense, the DNN can be seen as a planning pipeline, where each layer settles a bit of the endeavor before offering it to the accompanying, until finally, the last layer gives the yield. Extra hid layers can be added to a *multilayer perceptron* neural framework to make it increasingly significant. The extra covered layers are fathomed to recombine the depiction from prior layers and make new depictions at noteworthy degrees of thought. A satisfactorily tremendous single covered layer multilayer perceptron can be used to derive most limits. Extending the significance of the framework gives another game plan that requires fewer neurons and gets ready faster. Finally, including significance, it is a kind of valid progression (Figure 7.8).

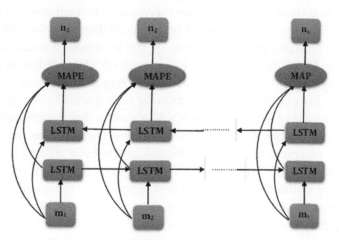

FIGURE 7.8   Stacked LSTM architecture.

Given that LSTMs take a shot at gathering data, it suggests that the alternative of layers incorporates levels of consideration of data recognitions after some time. As a result, lumping perceptions after some time or speaking to the issue at various time scales. This methodology conceivably permits the concealed state at each level to work at various timescale. It is the profundity of neural systems that is, for the most part credited to the achievement of the methodology on a wide scope of testing forecast issues. Stacked LSTMs are currently a steady system for testing arrangement expectation issues. A Stacked LSTM design can be characterized as an LSTM model involved different LSTM layers. An LSTM layer above gives an arrangement yield as opposed to a solitary worth yield to the LSTM layer underneath. In particular, one yield for every information time step, instead of one yield time step for all information time steps.

7.6.2   *BIDIRECTIONAL LSTM*

Bidirectional LSTMs [10] are a growth of traditional LSTMs that can impro-vise structure execution on the course of action modeling issues. In issues where unparalleled steps of the data game plan are available, Bidirectional LSTMs trains two LSTM instead of one LSTM on the data gathering. The first on the data gathering without any assurances and the second on a pivoted copy of the data course of action. This can give additional settings to the framework and result in speedier and essentially progressively full learning on the issue. The possibility of *bidirectional recurrent neural networks* (RNNs) is direct. It includes copying the principal intermittent layer in the system so that there are presently two layers next to each other, at that point giving the info arrangement as-will be as a contribution to the primary layer and giving a turned around duplicate of the information succession to the second. To defeat the restrictions of a normal RNN, a bidirectional-based RNN is suggested that can be prepared utilizing all accessible information data before and eventual fate of a particular time span. The thought is to part the state neurons of a customary RNN in a section that is answerable for the positive time heading (forward states) and a section for the negative time bearing (in reverse states). This methodology has been utilized to extraordi-nary impact with LSTM-based RNN frameworks. The utilization of giving the grouping bi-directionally was at first supported in the space of discourse acknowledgment on the grounds that there is proof that the setting of the entire expression is utilized to decipher what is being said instead of a straight translation. All things considered, timesteps in the information succession are as yet handled each in turn, it is only the system ventures through the information grouping in the two headings simultaneously (Figure 7.9).

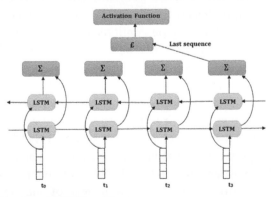

FIGURE 7.9   Bidirectional LSTM architecture.

7.6.3 STACKED GRU

Stacked GRU [10] can be prepared the succession information like by the stack and procedure the information individually. The stacked GRU can be characterized as an LSTM model contained numerous LSTM layers. Stacking the GRU layers in the middle of concealed layers can make the model further for precise investigation. This stacked GRU comprises various GRU layers in the middle of the info and the yield. The different layers of GRU have given the more profundity level of Precision in the GRU organization. The benefit of the stacked GRU is higher exactness in expectation of the drawn-out clutters (Figure 7.10).

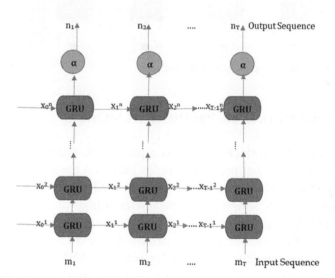

FIGURE 7.10   Stacked GRU architecture.

7.6.4 BIDIRECTIONAL GRU

Bi-directional GRU [10] is dependent on GRU, yet it performed both forward and in reverse. It associates two concealed layers of inverse bearings to a similar yield. The possibility of Bi-Directional GRUs gets from bidirectional RNN, which procedures request information in both forward and in reverse ways with two discrete shrouded layers. Bidirectional GRU appends the two concealed layers to a similar yield layer. It has been demonstrated that the bidirectional systems fundamentally superior to the unidirectional system (Figure 7.11).

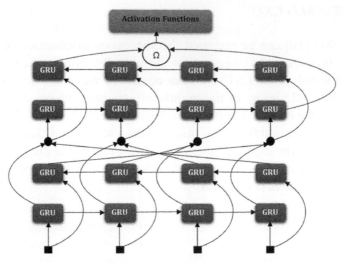

FIGURE 7.11   Bidirectional GRU architecture.

7.7   RNN-BASED DISEASE PREDICTION MODELS

Recurrent neural network (RNN) [6] is a part of DL methods, which is the first algorithm in DL that produces output based on the sequence of input. RNN has multiple advantages in the field of medicine to solve it. LSTM, an extension of RNN solves hyper-parameter setting problem in RNN by using to store the long sequence of memory through cell state. Predicting non-communicable diseases such as cardiovascular disease using the RNN techniques is predominant among older patients. ANN is to connect the nodes which mimic the neuron in the human brain. In ANN, the information is passed to every hidden layer to save the updated information. There are many Activation functions in the DL which is used to bring the efficient training and to get the accuracy of the prediction.

In the RNN, the main technique of GRU is useful for predicting heart disease. The ReLU activation function allows the network for the back-propagation to check which weight of the input gives the best accuracy for the prediction of the disease. To give the better optimization of the DL, the loss function and the optimizer techniques are used and checked between the RNN and convolutional neural network (CNN). Along with the activation function and the optimizer, the more number of hidden layers also gives the accuracy of the process of saving the information for a long time [9].

Using the LSTM and deep care, the diabetes and mental health are predicted. Diabetes is developed based upon the blood glucose level along with this blood sugar level the features of heart disease is found to predict heart disease and also the heart disease are found by Evolutional Rule Mining. In this chapter from the diabetes patients, the cardiovascular disease is prognosticated using the data mining methods. Recent research builds DL models based on the above features for classifying neurodegenerative disease. Recently, neurodegenerative disease identification was done based on CNN. CNN technique increases the complexity with less accuracy. Later, diagnosis is carried over based on the premature identification of Alzheimer's disease (AD) using deep neural networks (DNN). It does not describe much about any specific architecture in the neural network. A report has been generated based on the study of different machine learning algorithms for disease prediction. Some researchers opinioned that minimal GRUs used, does not achieves a satisfactory accuracy. Vanilla RNN, LSTM, GRU techniques are used for prediction, but Precision, recall, p-measure metric values is not accurate ones [8].

Scientists have as of late begun to apply profound learning techniques to clinical applications. Autoencoders are utilized to take in phenotypic examples from serum uric corrosive estimations. Profound neural systems with gradual learning on clinical time arrangement information have found physiologic examples related to known clinical phenotypes. Additionally, Boltzmann machines are applied confined on time arrangement information gathered from wearable sensors to anticipate the sickness province of Parkinson's malady patients. LSTM models are utilized for multi-name analysis expectation utilizing pediatric ICU time arrangement information (e.g., pulse, circulatory strain, glucose level, and so on.). Numerous examinations utilized multivariate time arrangement information from patients, which concentrated on totally different clinical conditions, with ceaseless time arrangement information. Many spotlights were on early discovery of HF for the general patient populace dependent on broadly accessible EHR information, for example, time-stepped codes (determination, prescription, procedure). DL methods have been as of late applied to clinical content information (e.g., PubMed abstracts, progress notes) utilizing Skipgram to learn connections among clinical procedures or bound together clinical language framework (UMLS) ideas. Numerous specialists applied Skipgram to longitudinal EHR information to learn low-dimensional portrayals for clinical ideas, for example, analysis codes, medicine codes, and method codes, and to learn portrayals of clinical ideas.

7.8   LSTM-BASED CARDIOVASCULAR DISEASES (CVDS) PREDICTION

Cardiovascular diseases (CVDs) are the main source of death overall. As per the WHO, about 17.9 million individuals kicked the bucket of CVD in 2016, representing 31% of everything being equal [11]. Arrhythmia is brought about by ill-advised intra-cardiac conduction or heartbeat development, which can influence heart shape or disturb the pulse. An electrocardiogram (ECG) is a far-reaching indication of the electrical sign action of the human heart. Acquiring the gritty physiological condition of different pieces of the heart by gathering signals is a vital method for clinical target determination. Mechanized investigation and finding dependent on ECG information have a solid clinical analytic reference an incentive for arrhythmia.

Numerous strategies for the programed characterization of ECGs have been proposed. The kind of ECG beat can be recognized when space, wavelet change, hereditary calculation, bolster vector machine (SVM), Bayesian, or different techniques [12]. In spite of the fact that the above order techniques accomplish high exactness on trial datasets, their exhibition is profoundly reliant on the extraction attributes of fixed or manual structure strategies. Physically planning extricated highlights may increment computational unpredictability all through the procedure, particularly in the changing area. Profound learning comprises the standard of AI and example acknowledgment. It gives a structure where highlight extraction and characterization are performed together. Profound learning has been generally utilized in numerous fields, for example, picture order, target discovery, and malady forecast.

LSTM organize an uncommon sort of RNN that is broadly utilized for time arrangement examination. It can adequately hold authentic data and acknowledge the learning of long haul reliance data of content. It has been utilized in numerous fields, for example, characteristic language preparing and discourse acknowledgment. Heart disease is one of the diseases that affect the human life more deeply. Many lives of humans are spoiled by heart problems. There are many reasons to get heart disease, but the patients of diabetes can get heart problems with the increase in the blood glucose level and other features. The risk factor such as high blood pressure and high cholesterol may cause a stroke or heart disease to diabetes patients. This heart disease gets predicted by DL techniques long short term memory (LSTM) and GRU. Heart disease is one of the major diseases among the peoples. The patients have chances to get a heart attack by the increasing level of blood glucose level. To predict this heart disease from diabetes, many surveys are carried over between LSTM and GRU techniques.

The prediction flow of the LSTM-based heart disease prediction framework is illustrated in Figure 7.12. Initially, the EHR clinical DB is preprocessed using a MinMaxScaler algorithm to remove the outlier and noisy data in the given dataset. After that, the input is passed onto the LSTM layer for DL prediction. The patient's events are converted into vector values which are then processed by the various hidden layers. These layers act as the activation window for prediction based on RELU, Sigmoid, and Tanh activation functions. At last MAPE algorithm is applied to check the performance of the derived model in terms of accuracy and regression.

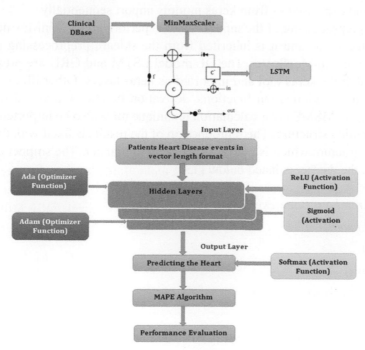

FIGURE 7.12   LSTM-based heart disease prediction framework.

7.8.1   *DIABETES PREDICTION FOR HEART DISEASE PATIENTS WITH LSTM*

The issue going to be executed is the diabetes expectation problem. This is where, given the cholesterol worth and coronary illness condition, the errand is to anticipate the event of diabetes for the patient in the future. The information utilized is Pima Indian gathered from Google Kaggle. The pima Indian dataset is altered with changes, for example, month to month report of the single patient for a year is planned with parameters specifically cholesterol worth and coronary illness condition.

Before the prediction coding process, import the entirety of the capacities and classes expected to use in the usage. This expects a working SciPy condition with the Keras profound learning library introduced. The MATLAB plotting library is imported as pyplot for plotting graphs of the predicted result. It is expressed as matplotlib import pyplot as plt. Here plt is used as the pyplot reference variable in the programming snippet. The sequential phase execution and Dense dataset processing is inherited from the keras model and expressed as from keras models import sequentially.

The preprocessing of the input dataset is performed by MinMax normalization technique and it is inherited from the sklearn preprocessing unit in the form of MinMaxScaler. The DL model LSTM and GRU are present in built within the keras tool and imported as keras.layers. Other library packages such as optimization functions, activation functions, tensor flow unit, NumPy, and MSME error calculation technique must also be imported in the programming structure. The reproduction of the results is fixed with the help of seed structure which is realized in a random manner. The snippet to load and plot the dataset is listed below [13]:

```
df = pd.read_csv('new.csv,' index_col='Patient id')
print(df.head())
df.plot()
plot.show()
columns_to_keep = ['Heart disease']
df = df[columns_to_keep]
df['Heart disease'] = df['Heart disease']
df['Cholestrol'] = df['Cholestrol']
df['Heart disease'] = df['Heart disease']
df.index.names = ['Patient id']
df.sort_index(inplace=True)
print('Total rows: {}.'format(len(df)))
df.head()
```

```
df.describe()
df.plot()
```

LSTMs are touchy to the size of the info information, explicitly when the sigmoid (default) or tanh actuation capacities are utilized. The dataset is then standardized utilizing the MinMaxScaler preprocessing class [13].

```
df.isnull().sum()
len(df[df['Heart disease'] == 0])
len(df[df['Cholestrol']==0])
print(np.min(df))
print(np.max(df))
ds = df.astype('float')
sca = MinMaxScaler(feature_range=(0, 3))
scal = sca.fit_transform(ds)
print(np.min(scal))
print(np.max(scal))
print(scaled[:10])
```

The next important step in prediction is partition of dataset into training part, testing part and validation part. Regularly, a ratio of 1:3 is followed for partitioning the dataset. Hence 70% of the data is utilized for the training phase, and 30% is realized for testing phase [13].

```
tr_s = int(len(scal) × 0.70)
te_s = len(scal-tr_s)
tr, te = scal[0:tr_s,:], scal[tr_s: len(scal),:]
print('train model: {}\n testing model: {}.'format(len(tr), len(te)))
def cr_da(ds, lk_bk=1):
print(len(ds), lk_bk)
dX, dY = [], []
for j in range(len(dataset)-lk_bk-1):
b = ds[j:(j+lk_bk), 0]
print(j)
```

```
print('P {} to {}.'format(j, j+lk_bk))
print(b)
print('Q {}.'format(j + lk_bk))
print(ds[j + lk_bk, 0])
ds[j + lk_bk, 0]
dX.append(b)
dY.append(ds[j + lk_bk, 0])
return np.array(dX), np.array(dY)
lk_bk = 1
```

The dataset creation method takes two arguments: first ds, which is a NumPy and the lk_bk, which is the quantity of past time steps to use as info factors to anticipate whenever period for this situation defaulted to 1 [13].

```
Ptr, Qtr = cr_dat(tr, lk_bk)
Pte, Qte = cr_da(te, lk_bk)
Ptr, Qtr = cr_da(tr, lk_bk)
Pte, Qte = cr_da(te, lk_bk)
```

The above method is utilized to set up the train and test datasets for demonstrating. To change the readied train and test input information into the normal structure, utilize np.reshape() as follows [13]:

```
Ptr = np.reshape(Xtr, (Xtr.shape[0], Xtr.shape[1], 1))
Pte = np.reshape(Xte, (Xte.shape[0], Xte.shape[1], 1))
print(Ptr.shape)
print(Pte.shape)
```

The sigmoid enactment work is utilized for the LSTM squares. The system is prepared for 100 ages and a clump size of 1 is utilized [13].

```
ba_si = 1
model = Sequential()
model.add(GRU(10, ba_in_sh=(ba_si, lk_bk, 1), stateful=True))
model.save_weights("record.h5")
```

```
model.add(Dense(6))
model.add(Dense(1))
model.compile(loss='mean_squared_error,'optimizer='adam,'metrics=['a
ccuracy'])
model.fit(Ptr, Qtr, epochs=4, ba_si=ba_si, verbose=2, shuffle=True)
tPredict = model.predict(X_train, ba_si=ba_si)
tPredict = model.predict(Pte, ba_si=ba_si)
```

Note that the expectations are reversed before computing blunder scores to guarantee that presentation is accounted for in indistinguishable units from the first information (a large number of new patients every month) [13].

```
trPre = scal.inverse_transform(trPret)
Qtr = scal.inverse_transform([Qtr])
tPredict = scaler.inverse_transform(tPredict)
Qte = scaler.inverse_transform([Qte])
trSc = math.sqrt(mean_squared_error(Qtr[0], tPredict[:,0]))
print('Training: %.3f RMSE'% (trSc))
teSc = math.sqrt(mean_squared_error(Qte[0], tPredict[:,0]))
print('Testing: %.3f RMSE'% (teSc))
tr_Pre_Plot = np.empty_like(scal)
tr_Pre_Plot [:,:] = np.nan
tr_Pre_Plot [lk_bk:len(tPredict)+lk_bk,:] = tPredict
```

At last, produce the forecasts model framework utilizing the trained and testing dataset [13].

```
# Plotting of predictions
te_Pre_Plt = np.empty_like(scal)
te_Pre_Plt [:,:] = np.nan
te_Pre_Plt [len(tPredict)+(lk_bk × 2)+1:len(scal)-1,:] = tPredict
pt.figure(fig_si=(30,20))
pt.plot(scaler.inverse_transform(scaled))
pt.plot(tr_Pre_Plt)
```

pt.plot(te_Pre_Plt)

pt.show()

7.9   RNN-BASED NEURODEGENERATIVE DISEASES PREDICTION

Neurodegenerative disease [14] involves the death of certain neuron cell in the human brain, causing problems such as ataxias and dementias. Such diseases are treated with the LSTM model, but accurate results are not able to achieve due to the gradient exploding problem. Neurodegenerative diseases destroy motor neurons in the central nervous system that wheels the gestures of limb movements. Machine learning technologies are efficiently smeared to the gait disorders in neurological disease. The deep neural systems are based on convolutional and recurrent architectures, which ascertain to be competent to course all forms of existing data and have more advantages when compared to other methods. LSTM is an extended method of RNN algorithm, which solves the vanishing and gradient problem in RNN, and it has a cell state to maintaining the long sequence of memory values. The BIGRU method was proposed with an extra layer added in Bi-stack GRU. This layer is directly attached to the output layer and can be useful to add the extra features to the dataset to predict future illness progression.

One promising model is the identification of neurodegenerative diseases through the patient's gait features using GRU. The disease progression can be detected easily, and it also increases the classification accuracy when compared to the LSTM model. Here Gaited Recurrent Unit in Figure 7.13, which is the extension of the LSTM model has an update gate and reset gate, these two functions are used to pass the information to the output and then trained to keep the long sequence of data without losing the previous information.

The update gate is used to determine how much information is going to pass along with the future value. The update gate is denoted by W(t) and it is calculated by using the formula as [15]:

$$W_t = \sigma \left(X^{(z)} y_t + V^{(z)} j_{t-1} \right) \tag{18}$$

Figure 7.13 depicts the overall working flow of the GRU. The sigmoid function (σ) having the added values of the previous state output j(t–1) and the current input y(t) along with the product of their weights and the results stored in w(t). The result of the sigmoid activation function lies between 0 and 1.

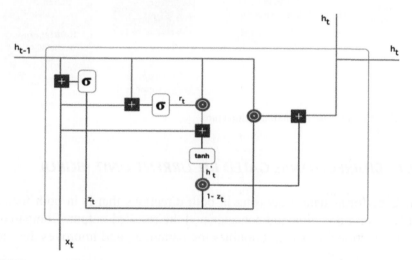

FIGURE 7.13   GRU Workflow.

Reset gate is used to decide how much of the past information need to be forgotten. It will take the same input as the update gate the difference takes in the weights and the gates. It is calculated by the formula [15]:

$$s_t = \sigma\,(X^{(r)}y_t + V^{(r)}j_{t-1}) \tag{19}$$

Hadamard product is applied on s(t) then applies the non-sequential tanh activation function to the input y(t) with s(t) and stores the result in j'(t). j'(t) and 1-z(t) combine to give the output of current state j(t). This can be articulated as follows [15]:

$$j'(t) = tanh\,(x(t) + (P(t)\,\Theta j(t{-}1))) \tag{20}$$

$$j_t = +\,(z_t\,\Theta j_{t-1}+(1{-}z_t)\,\Theta j't \tag{21}$$

In addition, the extension of the GRU is bidirectional gated recurrent unit (BGRU) and stacked GRU. In our case, the preprocessed data is given to the extension of GRU to get the result (Figure 7.14).

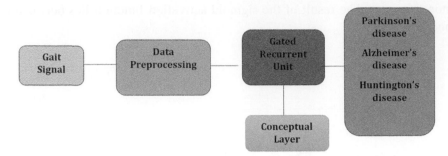

FIGURE 7.14   Disease prediction architecture.

7.9.1   *BIDIRECTIONAL GATED RECURRENT UNIT (BGRU)*

BGRU performs data processing in both directions that is in both forward and backward direction and concatenated the resulted output. Compared to the unidirectional process, it doubles the parameter and improves the accuracy of the process.

$$J(r) = (-\rightarrow j(r).\leftarrow - j(r)) \tag{22}$$

The dataset is processed using the BGRU initially and then apply the stacked gated recurrent unit (SGRU) for the multiple layers.

7.9.2   *STACKED BIDIRECTIONAL GATED RECURRENT UNIT (SGRU)*

SGRU increases the depth of the layers used in the GRU model to improve the accuracy. Here neurodegenerative dataset is processing in both forward as well as backward direction with increasing depth shows in Figure 7.15.

7.9.3   *CONTEXTUAL LAYER*

The contextual layer is an extra layer added to bi-stack GRU. This layer is directly attached to the output layer, and this layer can be useful to add the extra features to the dataset to predict future illness progression. Extra features like blood pressure and blood glucose level can be added it can be useful to predict the risk of diabetes in the future.

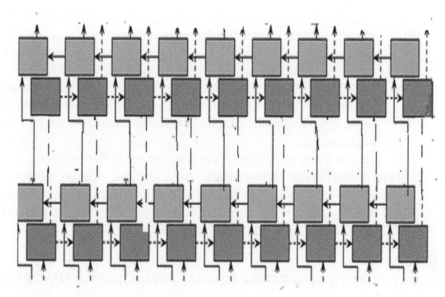

FIGURE 7.15   Stacked bidirectional GRU.

7.9.4   PREDICTION MODULES

1. **Data Collection:** It is a process of summarizing medical information collected from different sources. A real-time dataset is used for
   neurodegenerative disease classification analysis. For this disease
   classification, two types of signal datasets are used one is gait signal
   and brain signal. Gait signal [14] is used to outline the walking gesture
   of the patient and the brain signal used to describes the signal from
   the brain to various parts of the patient's body through the images.
   Sensor and electroencephalogram (EEG) is used for collecting the
   Gait signal and brain signal, respectively.

2. **Data Preprocessing:**
   a. **Noise Removal:** Data that is collected from the data collection
      having the possibilities of noisy data because of false apparatus
      can be removed using some noise filter in order to avoid the low
      accuracy. Data contains imbalanced, inaccurate, and contami-
      nated value can be preprocessed before the processing it should
      be removed to prevent the inaccurate results and efficiency.
   b. **Signal to EDF Conversion:** The dataset must be in the format
      of .xls or .csv or edf because the Keras library accepts this

format only. Keras contains a package to import the EDF file directly into the snippet. Many open-source tool is available for converting the EEG signal to ECG conversion. EDF browser is one of the best tools to convert EEG signals to EDF file.

c.   **EDF to CSV:** Once data is converted into EDF format, through Keras package it is imported directly to the snippet or converts into a CSV file which contains the separate column and row for every patient. The features like age, weight, gait speed, gender, etc., in a row and patient's diseases list are in the column.

3.   **Gated Recurrent Unit (GRU):** A DL model, and an advanced version of long short term memory (LSTM). In addition, it solves the hyper-parameter training problem in recurrent neural network (RNN) using an update gate and reset gate.

7.9.5   NEURODEGENERATIVE DISEASE PREDICTION USING RNN

The problem of going to be implemented is the neurodegenerative disease prediction. This is a problem where, given the age, BMI, blood glucose level, pregnancy counts, skin thickness, blood pressure, and insulin Level, the task is to predict the occurrence of the Neurodegenerative disease for the patient depicted through outcome parameter. The data used is Pima Indian collected from Google Kaggle with modified parameter for brain disease occurrence. The following programming process is common to both LSTM and GRU strategy.

The MATLAB plotting library is imported as a pyplot for plotting graphs of the predicted result. It is expressed as a matplotlib import pyplot as plt. Here plt is used as the pyplot reference variable in the programming snippet. The sequential phase execution and Dense dataset processing is inherited from the keras model and expressed as from keras.models import sequential.

The preprocessing of the input dataset is performed by the MinMax normalization technique, and it is inherited from the sklearn preprocessing unit in the form of MinMaxScaler. The DL model LSTM and GRU are presented inbuilt within the keras tool and imported as keras.layers [16].

Other library packages such as optimization functions, activation functions, tensor flow unit, numpy, and MSME error calculation technique must also be imported in the programming structure. The reproduction of the results is fixed with the help of seed structure which is realized in a random manner. For the splitting up of the dataset, the model selection package is imported from the Sci learning kit to phrase the training-testing split up. The snippet to load and plot the dataset is listed below [16]:

GRU and LSTM Model

 # Dataset-Loading of PIMA INDIAN
data = np.loadtxt('E:\Brain.csv,' delimiter=,," skiprows=1, skipcols=1)
data.plot()
plot.show()

LSTMs are delicate to the size of the information, explicitly when the activation functions enactment capacities are utilized. Then, rescale the information to the scope of 0-to-1, additionally called normalizing. Standardize the dataset utilizing the MinMax normalization for scaled values [16].

GRU and LSTM Model

df.isnull().sum()
len(df[df['Heart disease'] == 0])
#len(df[df['Cholestrol']==0])
#print(np.min(df))
#print(np.max(df))
ds = df.astype('float')
sca = MinMaxScaler(feature_range=(0, 3))
scal = sca.fit_transform(ds)
#print(np.min(scal))
#print(np.max(scal))
#print(scaled[:10])

GRU Model

 # Variables Splitting-I/P and O/P
P = data[:,0:8]
Q= data[:,7]

LSTM Model

 # Variables Splitting-I/P and O/P
P = data[:,0:7]
Q = data[:,6]

The following snippet splits the input dataset model into two partitions with the range of 60% for the training phase and 40% for the testing phase [16].

GRU Model

 # split the data into training (60%) and testing (40%)
(P_tr, P_te, Q_tr, Q_te) = tr_te_split(P, Q, te_siz=0.33, ran_st=seed)
P_tr = np.reshape(P_tr, (P_tr.shape[0], 1, P_tr.shape[1]))
Q_te = np.reshape(Q_te, (Q_te.shape[0], 1, Q_te.shape[1]))

LSTM model

(P_tr, P_te, Q_tr, Q_te) = tr_te_split(P, Q, te_siz=0.33, ran_st=seed)

The system has a noticeable layer with single info, a shrouded layer with four LSTM neurons, and a yield layer that makes a solitary worth forecast. The default sigmoid initiation work is utilized for the LSTM squares. The system is prepared for 100 ages and a clump size of 1 is utilized [16].

GRU Model

 # Model Creation
mod = Sequential()
mod.add(Dropout(0.2, input_shape=(8,)))
mod.add(GRU(20, input_dim=8, init='uniform,' activation='relu'))
mod.add(Dense(14, init='uniform,' activation='sigmoid'))
mod.add(Dense(7, init='uniform,' activation='sigmoid'))
mod.add(Dense(1, init='uniform,' activation='softmax'))

LSTM Model

mod = Sequential()

mod.add(LSTM(6, input_dim=7, init='uniform,' activation='sigmoid'))
mod.add(Dense(1, init='uniform,' activation='relu'))

The neuro disease prediction model is compiled with loss function-binary cross-entropy and Adadelta optimizer. The accuracy metric is used to evaluate the model performance [16].

model compilation
mod.compile(loss='binary_crossentropy,' optimizer='Adadelta,'
metrics=['accuracy'])

Finally, to generate the graphical representations of the predicted output in the context of neuro disease analysis can be performed utilizing the following snippet [16]:

GRU Model
 # Model fitting
mod.fit(P_tr, Q_tr, validation_data=(P_te, Q_te), nb_epoch=2, bat_si=5,
verbose=1)
 # Model evaluation
scores = model.evaluate(P_te, Q_te)
print("Accuracy Value: %.3f%%%" % (scores[1] × 100))

LSTM Model
P_tr = np.reshape(P_tr, (P_tr.shape[0], 1, P_tr.shape[1]))
P_te = np.reshape(P_te, (P_te.shape[0], 1, P_te.shape[1]))
 # Model fitting
mod.fit(P_tr, Q_tr, validation_data=(P_te, Q_te), nb_epoch=2, bat_si=5,
verbose=1)
 # Model evaluation
scores = model.evaluate(P_te, Q_te)
print("Accuracy Value: %.3f%%%" % (scores[1] ×100))

The following output will be produced as a result of program execution in the conda environment.

GRU Model

Epoch 1/2

*227/227 {**************}-2s 100ms/step-loss: 7.7056-acc: 0.5110-val_loss: 0.690-val_acc: 0.7658*

Epoch 2/2

*227/227 {**************}-0s 41.3ms/step-loss: 7.7956-acc: 0.5110-val_loss: 0.690-val_acc: 0.7658*

*117/117 {**************}-0s-Stop*

Accuracy: 76.58%

LSTM Model

Epoch 1/2

*227/227 {**************}-1s 100ms/step-loss: 160.4982-acc: 0.3286-val_loss: 140.1574-val_acc: 0.3478*

Epoch 2/2

*227/227 {**************}-0s 335ms/step-loss: 160.4982-acc: 0.3286-val_loss: 140.1574-val_acc: 3478*

*117/117 {**************}-0s-Stop*

Accuracy: 34.78%

7.10   CONCLUSION

AI is the study of the human intelligence in which machine learns the perspective functions that human-related with another human mind. Machine learning (ML) makes the software proficient in learning on the basis of their prior experience and improves the efficiency of the prediction task. DL is the subspace of the AI that consists of the neural network, which refers to the depth of the network architecture, going deeper towards the data performance in the task. Our contribution in this chapter relates to the role played by recurrent neural DL network architecture in the field of biomedical engineering and its applications. Among various RNN architecture, LSTM and GRU exhibited a growing interest among researchers, supplanting traditional machine learning algorithms. The exploration of this research showed that RNN structures, especially LSTM and GRU, have a greater ability in projecting the knowledge of diverse classification of heart and neuro diseases.

KEYWORDS

- **artificial neural network**
- **bidirectional gated recurrent unit**
- **deep learning**
- **discriminative learning**
- **generative learning**
- **long short term memory**
- **neuro-disease**

REFERENCES

1. Abdelrahman, H., & Anthony, P., (2010). A study of deep learning. *International Journal from the University of Connecticut.*

2. Liyan Xu. (2019). *Variants of RNN Models.* Available at: https://liyanxu.blog/2019/01/24/rnn-variants-gru-lstm/ (accessed on 18 December 2020).

3. Moher, D., Liberati, A., Tetzlaff, J., Altman, D. G. (2009). Preferred Reporting Items for Systematic Reviews and Meta-Analyses (PRISMA). Available at: http://www.prisma-statement.org/ (accessed on 18 December 2020).

4. Aditi Mittal (2019). *Understanding RNN.* Available at: https://towardsdatascience.com/understanding-rnn-and-lstm-f7cdf6dfc14e (accessed on 18 December 2020).

5. Sepp, H., & Jurgen, S., (1997). Long short-term memory. *Neural Computation, 9*(8), 1735–1780.

6. Kyunghyun, C., Bart, V. M., Caglar, G., Dzmitry, B., Fethi, B., Holger, S., & Yoshua, B., (2014). *Learning Phrase Representations using RNN Encoder-Decoder for Statistical Machine Translation.* EMNLP, available at: https://arxiv.org/pdf/1406.1078.pdf (accessed on 18 December 2020).

7. Bin, D., Huimin, Q., & Jun, Z., (2018). Activation Function and their characteristic in deep neural network. *The Chinese Control and Decision Conference (CCDC).*

8. Reddy, S. V. G., Thammi, R. K., & Vallikumari, V., (2018). Optimization of deep learning using various optimizers, loss function and dropout. *International Journal of Recent Technology and Engineering (ISRTE), 7.* ISSN: 2277-3878.

9. Sayrabh, K., (2012). Approximating number of hidden layer neurons in multiple hidden layer BPNN architecture. *International Journal of Engineering Trends and Technology, 3*(6).

10. Rahul Dey; Fathi M. Salem, (2017). Gate-variants of gated recurrent unit (GRU) neural networks. *IEEE 60th International Midwest Symposium on Circuits and Systems (MWSCAS).*

11. Aakash Chauhan; Aditya Jain; Purushottam Sharma; Vikas Deep, (2018). Heart disease prediction using evolutionary rule learning. *International Conference on Computational Intelligence and Communication Technology (CICT).*

12. Ammar, A., & Al-Moosa, A. A. A., (2018). Using data mining techniques to predict diabetes and heart disease. In: *2018 4ᵗʰ International Conference on Frontiers of Signal Processing.*

13. https://machinelearningmastery.com/stacked-long-short-term-memory-networks/ (accessed on 18 December 2020).

14. Aite Zhaoa , Lin Qia , Jie Lia , Junyu Donga, & Hui Yub, (2018). *LSTM for Diagnosis of Neurodegenerative Diseases Using Gait Data.* Department of computer science and technology. Ocean University of China, Qingdao, China University of Portsmouth, Portsmouth, UK.

15. *Understanding GRU Networks.* Available at: https://towardsdatascience.com/illustrated-guide-to-lstms-and-gru-s-a-step-by-step-explanation-44e9eb85bf21 (accessed on 18 December 2020).

16. Phi, M. (2018). *Illustrated Guide to LSTM's and GRU's: A Step by Step Explanation.* Available at: https://towardsdatascience.com/illustrated-guide-to-lstms-and-gru-s-a-step-by-step-explanation-44e9eb85bf21 (accessed on 18 December 2020).

CHAPTER 8

MEDICAL IMAGE CLASSIFICATION AND MANIFOLD DISEASE IDENTIFICATION THROUGH CONVOLUTIONAL NEURAL NETWORKS: A RESEARCH PERSPECTIVE

K. SURESH KUMAR,[1] A. S. RADHAMANI,[2] S. SUNDARESAN,[3] and T. ANANTH KUMAR[4]

[1]Associate Professor, Department of Electronics and Communication Engineering, IFET College of Engineering, Tamil Nadu, India, E-mail: sureshkumarkskphd@gmail.com

[2]Professor, Department of Electronics and Communication Engineering, V. V. College of Engineering, Tamil Nadu, India, E-mail: asradhamani@gmail.com

[3]Assistant Professor, Department of Electronics and Communication Engineering, SRM TRP Engineering College, Tamil Nadu, India, E-mail: sundaresanece91@gmail.com

[4]Assistant Professor, Department of Computer Science and Engineering, IFET College of Engineering, Tamil Nadu, India, E-mail: ananth.eec@gmail.com

ABSTRACT

The medical image classification assumes a basic job in clinical treatment and instructing aid. The old approach about this medical image classification might have touched its maximum level considering the execution. Additionally, by utilizing them, much time and exertion should be spent on extricating and choosing characterization highlights. The profound neural system is a developing AI technique that has demonstrated its potential for various order assignments. Strikingly, the convolutional neural system overwhelms with the best outcomes on shifting picture-grouping assignments. Notwithstanding, clinical picture datasets are difficult to gather since it needs a ton of expert mastery to name them. In this chapter, a comprehensive analysis of various approaches to the medical image classification using convolutional neural networks (CNN) is presented. Here a short-term explanation of numerous datasets of medical images along with the approaches for facilitating the major diseases with CNN is discussed. All current progress in the image classification using CNN is analyzed and discoursed. Adding a feather to the cap, research-oriented points are presented for medical image classifications and identification of diseases which arises to humankind. This assessment could furnish completely the medical examination networks through the important information on the way to ace the idea of CNN in order to use it intended for refining the general humanoid social protection framework.

8.1 INTRODUCTION

Medical image classification is the most significant problems in the picture acknowledgment territory, and its point is to characterize clinical pictures into various classifications to help specialists in malady determination or further research. In general, clinical picture characterization can be partitioned into two stages. The initial step is extricating viable highlights from the picture. The subsequent advance is utilizing the highlights to construct models that group the picture dataset. Formerly, specialists generally utilized their expert experience to extricate highlights to order the clinical pictures into various classes, which is normally a troublesome, exhausting, and tedious errand. This methodology is inclined to prompting unsteadiness or no repeatable results. Thinking about the examination as of not long ago, clinical picture characterization application inquire and has had incredible legitimacy. The analyst's endeavors have prompted an enormous number of distributed examinations. In any case, at present, we, despite everything, can't achieve this crucial. On the off chance that we could complete the arrangement work fantastically, at that point, the outcomes would assist clinical specialists with diagnosing illnesses with further examination. Thusly, how to viably tackle this assignment is vital [41].

In a classification of image-related issues, the elucidation, and discriminative intensity of highlights extricated are basic to accomplish great order execution. Artificial neural network (ANN) has been read for a quite long time and takes care of complex arrangement issues including picture grouping [30]. The particular favorable position of neural system is that the calculation could be summed up to understand extraordinary sorts of issues utilizing comparative plans. Convolutional neural system (CNN) is an effective case of endeavors to demonstrate well-evolved creature graphical cortex utilizing ANNs. The design of CNN consumes solid natural conceivable proof help from Hubel what's more; Wiesel's premature effort preceding the feline's graphic cortex [31]. It partakes exhibited unrivaled execution in comprehending some hard picture order issues.

In specific applications, for example, the recognition of movement sign, framework-using CNN might even outperformed anthropological ability in benchmarking assessments [32]. The ultimate goal is to adopt the convolutional neural networks (CNNs) for classifying the various types of patterns of lung images, chest images, tomography images, MRI and CT images, etc. More than 150 disorders of lungs is been identified by using the CNN approach [33]. This process will cause some scarring effect on the lung tissues, and this may cause some difficulties in breathing. HRCT (high-resolution computed tomography) imaging remains used in differentiating the various types of lung disease patterns. Because of variation in visualization of the same set of images and the similarities within the different cases is a difficult task for achieving accuracy in classification, a modified CNN network is suggested for the identification of any kind of diseases by classifying the image pattern obtained from various means. It uses a supervised algorithm for the classification; hence the expected performance is achieved. Some of the other issues related to the modeling is been reduced by using the customized model.

CNN is the unique DL strategies that are used to take care of PC vision and examination issues. The principal commitment of this investigation is to bring utilization of CNN strategy into the front line in regards to divided comet measure pictures. The investigation evaluates and group's portioned comet examines pictures into four distinctive harm levels by CNN, which stands as a novel technique. The examination is composed as follows. In the main area, convention, and utilized materials were referenced to perform research facility tests. In the subsequent segment, the information securing process and proposed CNN model were clarified. In the third segment, results were introduced. In the fourth area, got results were examined. In the fifth area, the end was advanced [10]. The utilization of the customary AI strategies, for example, bolster vector techniques (SVMs), in clinical picture grouping started quite a while in the past. Be that as it may, these techniques have the accompanying disserves: the exhibition is a long way from the common sense norm and creating of them is much delayed lately. Likewise, the component extricating and determination are tedious and fluctuate as indicated by various items [34]. The profound neural systems (DNN), particularly the convolutional neural systems (CNNs) are generally utilized in changing picture order errands and have accomplished huge execution since 2012 [35]. The clinical pictures are difficult to gather as the gathering and marking of clinical information stood up to with the two information protection concerns and the necessity for tedious master clarifications. In the two general settling headings, one is to gather more information, for example, publicly supporting or diving addicted to the current medical information [36].

One more way is concentrating in what manner to expand the presentation of a little dataset that might be significant in light of the fact that the information accomplished from the examination can relocate to the exploration on large datasets. Another way is concentrating on how to expand the exhibition of a little dataset, which is significant in light of the fact that the information accomplished from the examination can move to the exploration on enormous datasets. What's more, the most huge distributed chest X-beam picture collection of data (Chest X-beam 14) be there unmoving far littler than the greatest general picture dataset-Image Net that could arrived at 14,197,124 occurrences while analyzing 10 years before [37, 38].

CNN-based techniques ought to adopt different methodologies to build the presentation of picture arrangement on little datasets: Some strategy is data augmentation [39]. It is based on reducing the size of training limits thus by augmenting the capability of the objective gratefulness model. Here the Gabor channel is utilized, and this is a lot of comparative in recurrence and its bearing through the human visual framework. The Gabor channel is near in repeat and course with the human visual structure. It is sensitive to the edge of the target and can give extraordinary course assurance characteristics. It is definitely not hard to evacuate information of the target at multi-scale and multi-course with Gabor channel.

8.2 CLASSIFICATION OF MEDICAL IMAGE USING CNN

Medical image classification comes under the topic image classification. Numerous methods in the classification of images that can likewise be utilized on it. For example, many picture upgraded techniques to improve the discriminable highlights for grouping [42]. A CNN remains substantiated by means of a most popular tool intended for the processing and resolving of classification problems in an image. Notwithstanding, as CNN is a start to finish answer for picture grouping, it become familiar with the component without anyone else. Along these lines, the writing about how to choose and upgrade includes in the clinical picture won't be looked into. The audit essentially centers on the utilization of conventional techniques and CNN based exchange learning. What's more, on the case organize on clinical picture associated document to examine anything influences in those representations are fundamental to the conclusive outcome and the holes might have excluded from the work.

Many researchers have been proposed in the name of CNN, that a vast improvement is identified for the better performance in view of a large number of databases (DBs) carrying image. Some of the image DBs are "modified national institute of standards and technology database (MNIST)," "NYU object recognition benchmark database (NORB)," and "Canadian institute for advanced research database (CIFAR10)" [1]. The learning of image data from a global level must lead to a positive approach in the field of research. Small range images like faces of human and the written images will have clear structures and so these small features can joined together to form a complex features. The deep CNNs are used here [40].

8.2.1 CONVOLUTION NEURAL NETWORK (CNN) FOR KNEE CARTILAGE SEGMENTATION

CNNs are used for the medical image analysis like for knee cartilage segmentation [2]. It is an automated segmentation process that is needed for showing improvement in the performance of Knee Osteoarthritis (OA) assessment, as the image is in 3D convoluted structure [3]. This approach uses CNN mixed with multiclass loss function. This is the best approach when compared to the SKI10 (a workshop held at china, MICCAI, in the year 2010) [4]. In this, the qualitative assessment that was obtained at the initial stage of segmentation yields the result that depicts visually about the cartilage loss in the longitudinal knee data obtained from the MRI images. Here another new system for classifying the voxel that integrates some 2D CNNs is implemented in association with the same MRI images, but the images taken are 2D images. These might have the association with the three different planes of 3D image (XY, ZX, and YZ planes). A segmentation process is done for the tibial cartilage image obtained from the MRI scans. The 2D features alone are used in this method. Compared to 3D multiscale features, it performs better [2].

8.2.1.1   AUTOMATED CARTILAGE SEGMENTATION

The MRI scan is a better choice for analyzing the thickness of the knee cartilage [5]. However, the information which is calculated manually about the cartilage thickness using the MRI scan is quite a difficult task and that may lead to severe error. This will make the regular analysis of individual or groups of patient's therapy planning to be impractical [6]. As this method integrates both 2D and 3D CNNs to reach an accurate and vigorous pathological knee shapes. Hence an automated cartilage segmentation method is used to explicate the relationship of complex images spatially and helps to successfully segment the cartilage [7]. Here an authenticated segmentation model that practices machine-learning algorithm is used for identifying the pixels of Magnetic resonance images which is of tissue type [8]. The CNN used in this approach is used for converting the grayscale assessment of every pixel corresponding to the radiograph images to not less than six numbers, and each represents six various tissue types. In this the CNN model used is U-net that already proves better results for the medical image segmentation [9].

8.2.2   DNA DAMAGE CLASSIFICATION USING CNN

DNA is the complicated part of the human physiology and the damage is highly crucial. Proper treatment is needed for this issue. Hence the DNA damage analysis is needed. The DNA image identification and classification is the prominent topic in the research field of biomedical engineering that require most effective methods, as it is one of the easy and the cheapest way to analyze. An approach which uses CNNs is implemented to train the test dataset. The augmented data set consists of more than 10,000 images belongs to either of the declared classes. No pre-processing parameters are required for the classification of DNA images and can classify the images into the declared classes (healthy, poorly defective, defective, more defective). The accuracy obtained using the above technique is greater than 96%. The CNN uses a method that extracts the learning features by means of filters through a neural network. Here the architecture of CNN has three layers (convolution, pooling, and classification). It accomplishes the extraction of features by means of filter matrices [10]. Both ANN and CNN are being used for the feature extraction without affecting the anomalies which were appear on the various regions of the image. CNNs reduce the usage of parameters cast-off for the learning purpose [11]. It uses MLP algorithm for the classification and the neurons found on the final layers must be moved to the vector region with the help of flattering. Thus the classification is done by the combination process of neurons and the hidden layers. Finally, the classification of images is done with the help of filters and neurons of Multilayer perceptron.

8.2.3 SOUND SEQUENCE CLASSIFICATION USING CNN

The CNN are intended for the ordering of the infant's vocal orders. The characteristics of the infant's normal sequences like crying, fussing, babbling, laughing, and other normal vocalization sequences are targeted for this analysis. The general instance of this characterization assignment remains existed for applications that could have need of a subjective assessment of a common new-born child vocalizations viz., torment evaluation or appraisal of language acquisition. The arrangement system depended on speaking to sound portions as spectrograms, which are contribution to a traditional CNN engineering plan. The impact of system is most efficiently investigated that includes the order execution to determine the rules for structuring viable CNN models for the task. CNN's ought to be demonstrated to have a little bottleneck between the convolutional organize and the completely associated arrangements, accomplishing through expansive accumulation of convolutional highlight maps over the period in addition to recurrence hub. The pre-eminent accomplishment of CNN setup produces a decent precision approx. 72% [12]. In another approach, the deep learning (DL) might be designed for the arrangement of sounds that occur in the environment using the spectrograms that gets generated from those sounds. The generated spectrogram image is obtained from the environmental sources and that is trained using CNN in addition to the TDSN (tensor deep stacking network). The data sets are analyzed, and after training, the accuracy achieved is about 49% while using CNN [13]. In this, the training section is conducted by enabling two activation functions they are Tanh and ReLU. It is just equal to the logical sigmoid activation function.

8.2.4   SKIN LESIONS CLASSIFICATION USING CNN

The most common disease occur in human is skin lesions. It will sometimes cause death to the patients. The grouping precision of skin injuries is a vital determinant of the achievement pace of relieving deadly maladies. Profound CNNs are presently the most pervasive PC calculations with the end goal of sickness order [17]. Deep CNNs designed for this learning is Inception-v3. The main use of this learning is training speed and transfer learning is considered to be more efficient. It consists of several millions of images accumulated dataset [18]. This inception v3 is used for many other complicated cases like cancer and diabetic-related treatment [19]. This model is utilized to contemplate the impact of picture noise on the order of skin injuries. All types of noises which occur in medical images (CT scan images, MRI scan images) are added to a picture dataset, to be specific the Dermofit Image Library from the University of Edinburgh. Assessments in light of t-conveyed Stochastic Neighbor Embedding (t-SNE) perception, receiver operating characteristic (ROC) examination, and saliency maps exhibit the dependability of the Inception-v3 profound CNN in ordering boisterous skin injury pictures.

8.2.5   ABNORMALITY DETECTION IN WCE IMAGES USING CNN

Color along with smoothness is considered as the major distinguishing features for detecting the abnormality in the WCE (wireless capsule endoscopy) images. The capsule endoscopy (CE) is the method to identify the pathologies position accurately [14]. The features of texture information and the color information is generated, and a CNN method might be applicable for retrieving the texture and color image data from the components obtained from the CIELAB color space (International Commission on Illumination). The patch in the edge remain appropriate to an anomaly is being identified by the defined algorithm by means of classification [15]. The Pre-processing should be done by bringing the overall dataset of images to the CIELAB. The images are then divided into various patches (approx. 100 nos per image). If there are 342 images, then there are about 34200 patches available. The CNN accepts the inputs in the size of (36×36×3) (height × width × channels). This is the normalized patch. The images are then padded and the patches are hauling out from the padded images. The normalization of the images will bring all the pixels to equal scale, and thus the convergence gets improved. The kernel is used for all layers, and the max-pooling (sample-based discretization process) layers is used.

The local response normalization (LRN) is castoff, and it provides the input for the max-pooling layer. It promptness up by doing the convergence process and the responses get damped. The ReLU is employed in all the layers. ReLU is having some needed properties which don't require input normalization that helps to prevent saturation. Some of the learning data's from the examples driven would produce an input that is provided as positive input to the ReLU [16]. Here multilayer perceptron is utilized.

8.2.6 TUBERCULOSIS (TB) IDENTIFICATION USING CONVOLUTIONAL NEURAL NETWORKS (CNN)

The TB is identified by using CNNs (a Bayesian-based approach). The uncertain cases are dealt with low perceptibility among the TB affected patients and others. The CNN with DL concept is used appropriately. Here two data sets are used, they are Shenzhen and Montgomery. These datasets are set as the benchmark, the testing and training is done using those dataset. The image is obtained from the chest x-rays (CXRs). Google colab platform is used for the purpose of training and testing [20]. The CNN are trained earlier to study the specific features from the available images sources obtained from the CXR. A DL method is used for improving the learning skills by enhancing the transfer of knowledge obtained for analyzing and detecting the TB [21]. The classification of the chest images are analyzed by using one stage of CNNs. Here a large number of networks is being used, and some of the layers are accompanied with the ReLU layers. However, at the same time, some layers have no max-pooling process. The chest array with some sets of data's are used to train the networks. The CXR 14 may consists of images with 14 types of disease information [28]. Of these, some are healthy images and some are undiagnosed images. From the overall images, some are unstable, and a number of identified patients obligate to numerous diseases. Hence to reduce the multi-class issues related to classification, the data set is being used to identify diseases like pneumonia. This model-based learning approach is used here for detecting TB from the available data collection. From this approach, the accuracy is obtained as 95% and the area under the curve is obtained as 95%, then the size of the chest images are obtained by using the rest of the layers in the CNNs.

8.2.7   BREAST CANCER LESION CLASSIFICATION USING CNN

The CNN consisting of multiple layers of connections related to neural computing has an insignificant range of methodical processing. Thus, yielding a substantial development for computer vision's region. As familiarly known, CNN architecture consists of some layers like convolutional, pooling, and completely connected which aim is to detect the edges, visual elements, and lines. The convolution involves the setting of some special filters which are learned through the system.

Generally, doctors need to physically depict the speculated bosom disease region. Various investigations have referenced that manual division requires some serious energy and relies upon the administrator. CNNI-BCC (CNN improvement for breast cancer classification) algorithm is implemented for assisting the medical specialists for diagnosing the breast cancer accurately in a short period of time. It has a CNNs structure which improves the classification capacity of breast cancer laceration for helping the experts for diagnosing the breast cancer [44]. This can classify the medically obtained cancer images into benign, malignant, and the normal healthiest patients. This algorithm can classify the images into all three categories and thus helps the experts for classifying the cancer lesion by the implementation of CNN. CNNI-BCC technique is introduced to help clinical specialists in bosom malignancy conclusion. Customarily, clinical specialists should physically depict the presumed bosom malignancy territory. Studies recommended that manual division requires some investment, yet additionally depends on the administrator [43]. CNNI-BCC is intended to execute a regulated profound learning neural system for bosom malignant growth characterization. The introduced work is an endeavor to help clinical specialists in deciding bosom malignant growth sore. The result shows that, the impending of CNNI-BCC indicates for analyzing the various circumstances of mistrusted breast cancer patients throughout analysis. A multiscale CNN is developed to categorize the breast images into few categories. They are benign lesion, normal tissue, invasive carcinoma, *in situ* carcinoma. The image is converted into the training patches and multiple scales are collected and are pre-trained. The accuracy obtained is more than 90% for the testing of 100 images [45]. This calculation could have been assessed preceding the BACH Challenge dataset, in addition to accomplish an exactness of 92% on the five-overlay cross approval on preparing pictures and an exactness of 90.04%.

8.2.8 PATCH BASED CLASSIFICATION USING CNN

Convolution neural network (CNN) in combination with the shallow convolution layer is used for the classification of image patches which were taken from the lung image for the person affected with interstitial lung disease (ILD). In this an CNN framework is designed with a customized manner, and it is then subjected to efficiently understand the image features inherently from the patches of lung image, which is more appropriate for the classification [1].

8.2.8.1 BREAST CANCER BY PATCH BASED CLASSIFICATION

A convolutional neural system alongside patch-based classifier (PBC) is produced for the grouping of pictures consequently from bosom malignancy histopathology images and divides into four above said classes. The PBC make use of the architecture of CNN for automatic classification. This suggested method will work into two modes: SPSD and MPSD (single patch in single decision and multiple patches in single decision). Malignant and non-malignant are the two classes that are being classified from the obtained histopathological images. It is again classified as into four subclasses, namely invasive carcinoma, normal, benign, and *in situ*. All the methods existed will have the classification as two classes. The accuracy achieved is about 87% by classifying the images from the test data set [47].

8.2.9 TOMOGRAPHY IMAGES CLASSIFICATION

Computerized arrangement of renal masses recognized at processed tomography (CT) assessments into generous blister versus strong mass is clinically important. This differentiation might be trying at single-stage differentiate improved CECT (contrast-enhanced computed tomography) assessments, where pimples may reproduce strong masses and where renal masses are most ordinarily by chance recognized. This may prompt superfluous and exorbitant follow-up imaging for precise portrayal [22]. A fix based CNN strategy to separate generous sores from strong renal masses utilizing single-stage CECT pictures. The expectations of the system for patches removed from a physically fragmented injury are consolidated through the lion's share, casting a ballot framework for conclusive analysis. A layer-based approach is being used here. The system was prepared utilizing patches that separated from preparing CECT checks and misleadingly expanded those outcomes in a sum of 65,287 patches (32,837 and 32,450 patches marked as sore and strong renal mass individually) for preparing. The formula for calculating the accuracy, precision, and the recall is given below:

$$\text{Accuracy} = \frac{P^T + N^T}{P^T + P^F + N^T + N^F}$$

$$\text{Precision} = \frac{P^T}{P^T + P^F}$$

$$\text{Recall} = \frac{P^T}{P^T + N^F}$$

where; P^T means true positive, N^T means true negative, P^F means false positive, and N^F means false negative. The solid renal masses and the cyst data is being obtained and is differentiated, which may yield the accuracy, precision, and recall as 89%, 96%, and 92%, respectively. The region of convergence curve is obtained, and it displays the performance of the proposed approach where the sensitivity and specificity are plotted for reference.

N = 327	Predicted Cyst	Predicted Solid
Cyst	27	12
Solid	24	264

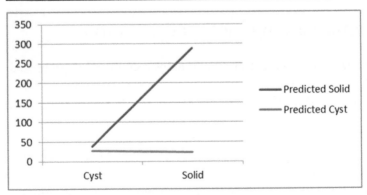

The consequences of investigation uncovered that robotized finding of amiable pimple remains more moving than strong tumors. The exhibition of CNN for growth cases is inspected and contrasted the outcomes.

Conventional technique, where generous blisters are separated from strong tumors dependent on the HU limit. Another novel approach for the identification of tumors is the usage of CNNs along with a DL-based approach. The algorithm used is the particle swarm optimization (PSO) algorithm. This classifies the various varieties of tumors (specifically the cancer based tumor) using three phases. Initially, the preprocessing method is used to select some of the optimal features by using the decision tree based PSO algorithm and convert the obtained images to 2D image. Next method is the process of enhancing the original dataset to be much larger by the process of augmentation. This method helps to reduce the overfitting related issues and the model is trained to provide the accuracy better. Finally, the trained image dataset is fed into the layers of CNN for extracting the features [23]. This method can detect all kinds of cancers, and the overall accuracy is obtained as 97%. Another method with the same DL mechanisms is used, that has a dataset that contains some images for the identification of intermediate regions. The result suggests that the accuracy for detecting the tumor region is notified as 99.15%. This is one of the fastest models for the determination of the tumors [29].

8.2.10 ALZHEIMER'S DISEASE (AD) CLASSIFICATION USING CNN

The graph theory is the oldest known method for the determination of certain parameters in the complex units such as brain activities. CNN's accompanied with the Graph models may help to obtain a certain representation of data's and it may intend for the organization of brain informations. Alzheimer's disease (AD) might be a neural disorder. It will cause the poor performance in the network functions. Therefore, graph-based CNN will be used for the classification of AD gamut into some categories like initial insignificant intellectual damage, cognitively normal, late slight intellectual weakening and AD [24]. The network is validated and trained by using the graph functions by diffusing the image data. The graph-based classifier performs better compared to the support vector machine (SVM) classifier based on the categories of the disease. Figure 8.1 shows the implementation of graphical CNN model.

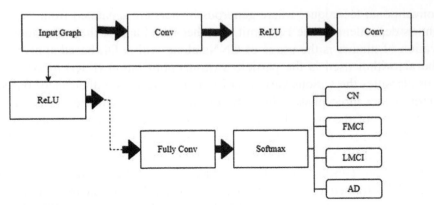

FIGURE 8.1 Architecture of the graph-based CNN.

The data needed for the training and study is obtained from the overall dataset for clinical data's and some human sample datasets available in Alzheimer's disease neuroimaging initiative (ADNI). It is a communal element which consumes a large quantity of datasets [25]. For augmenting the data's the synthetic sampling (adaptive) is casted-off [26], and this overcomes some of the drawbacks identified from another method called SMOTE (synthetic minority oversampling technique). This is an extension of ADASYN (adaptive synthetic sampling) method. It will generate the minority samples needed for training. From the whole dataset, half a portion is used for training, and another half is used for validation [27]. For validating or evaluating the performance of the classifier, a ROC curve is obtained where the sensitivity and specificity is determined. Several ROCs are computed, and the overall accuracy is obtained.

A multi-class GCNN (Graph CNN) classifier have actualized and approved for characterization of subjects on the AD range. The system utilizing basic availability diagrams is prepared and approved that is dependent on DTI (diffusion tensor imaging). ROC investigation results show that the GCNN classifier beats SVM by edges that are dependent on sickness classification. Discoveries show that the exhibition hole between the two techniques increments with sickness movement from CN to AD. The outcome is significant from a specialized viewpoint, since it clarifies the potential for the GCNN classifier to yield superior under low-example size settings. The underlying usage depends on basic L2 misfortune work. As future work, this classifier can be stretched out to bigger datasets, test elective misfortune capacities and test different GCNN designs and executions. A multi-model profound learning system is proposed, which dependent on CNN for joint programmed hippocampal division and AD arrangement utilizing basic MRI information. DL structure comprises of two profound learning models. One model is to perform multiple tasks profound CNNs for together learning hippocampus division and infection order, which creates a double division veil of the hippocampus and learns highlights for sickness characterization. In any case, the scholarly highlights that perform multiple tasks model are not adequate for exact sickness arrangement. A 3D fix covering the hippocampus is extricated dependent on the centroid of the division veil and contribution to a 3D DenseNet model that learns progressively applicable highlights for illness order. At long last, a completely associated layer and a SoftMax layer are affixed to join which took in highlights from these models for definite malady order. In this investigation, two classes are considered in the order task. The contribution of this system is an enormous picture fix covering the hippocampus. The yields are the hippocampus cover just as the expectation of illness status [46].

8.2.11 LIVER LESION CLASSIFICATION USING CNN

DL strategies and specifically convolutional neural systems (CNNs) have prompted a gigantic achievement in a wide scope of PC vision assignments, essentially by utilizing huge scope clarified datasets. Be that as it may, acquiring such datasets in the clinical space stays a test. A technique for creating engineered clinical pictures is presented that utilizes as of late introduced profound learning generative adversarial networks (GANs). Moreover, it is that the created clinical pictures can be utilized for engineered information increase, in addition to the improvement in the exhibition of CNN for clinical picture arrangement. The epic technique is on a constrained dataset of figured tomography (CT) pictures of 182 liver injuries (53 pimples, 64 metastases, and 65 hemangiomas). Initially, GAN designs for integrating top-notch liver sore ROIs are endeavored. At that point, a new plan is open for liver sore characterization utilizing CNN. At last, the CNN is trained for utilizing great information enlargement and engineered information growth and execution. Likewise, investigation of the nature of orchestrated models utilizing perception and master appraisal. The order execution utilizing just great information enlargement yielded 78.6% affectability and 88.4% explicitness. By including the engineered information growth the outcomes expanded to 85.7% affectability and 92.4% explicitness. This way to deal with engineered information expansion can expect to sum up to other clinical grouping applications and, in this manner, bolster radiologists' endeavors to improve diagnosis [48]. This concentrates on creating engineered clinical pictures with GAN for information increase to amplify little datasets and improve execution on grouping utilizing CNN. Moderately little dataset emulates the scope of datasets accessible towards most specialists in the clinical imaging network (by difference to the PC vision network somewhere huge scope datasets remain accessible).

8.2.12   BRAIN TUMOR CLASSIFICATION USING CNN

Tumors that occur in the human brain might be a compilation of typical cells. This prompts increment in the decease proportion amongst the people. A combination technique is suggested for consolidating the auxiliary in addition with the surface data of some MRI image successions (Flair, T1C, T1, and T2) designed for the location of brain tumor. The change in discrete wavelet alongside Daubechies wavelet portion is used for combination procedure that might give a progressively enlightening tumor region when contrasted with a specific solitary arrangement of MRI images. Subsequently the combination procedure, an incomplete differential dissemination channel (PDDF) is useful to expel clamor. A worldwide threshold strategy is utilized for dividing tumor region that could be then taken care of and to proposed CNN model for at long last separating tumor and non-tumor areas. Some freely accessible datasets available are BRATS 2012, BRATS 2013, BRATS 2015, BRATS 2013 Leader board, and BRATS 2018 which are utilized for projected technique assessment. The outcomes illustrates that the combined pictures give better outcomes when contrasted with singular successions on benchmark datasets. The strategy contains four significant advances. Initially, the four groupings of MRI might be intertwined into a solitary picture. In addition, PDDF is utilized for clamor expulsion, and in the third step; worldwide thresholding procedure could be applied for division. Consequential fragmented pictures will be assumed to recommended the CNN model in the last advance. Model utilizing convolutional, bunch standardization, ReLU, max pooling, completely associated and SoftMax layers to arrange non-tumor in addition to tumor pictures [49].

8.2.13   CT BRAIN IMAGE CLASSIFICATION

Recently, imaging might developed to a basic segment in numerous fields of medical research. Investigation of the various medical image categories might require modern representation in addition to preparing apparatuses. Deep neural systems obligate to present themselves by means of some most significant parts of AI and have remained effectively utilized in numerous fields for acknowledgment and clinical imaging applications. Amongst the various systems, CNN that could be organically motivated variations of multilayer observations are broadly utilized in the clinical imaging arena. In these systems, enactment work assumes a noteworthy job, particularly when the information comes in various scales. There is a hunger to improve the presentation of these systems by utilizing versatile initiation capacities which adjusts their parameters to the information. Here a changed rendition of a fruitful convolutional neural system is utilized for tuning medical image characterization furthermore, examined the impact of applying three kinds of versatile actuation works. These enactment capacities join essential actuation works in straight (blended) and nonlinear (gated what's more, various leveled) ways. The adequacy of utilizing these versatile capacities is appeared on a CT brain images dataset (as like complex clinical dataset). The trials demonstrates that the order precision of the projected is connected by means of versatile enactment capacities stands sophisticated when compared with the ones utilizing essential initiation capacities [52].

8.2.14   TYMPANIC MEMBRANE CLASSIFICATION USING CNN

The tympanic membrane is inspected clinically by the experts and conducted visually. This inspection indicates to a limited unpredictability amongst the spectators and the errors induced by the humans. Here these problems are solved by using a new model which is based on a faster regional CNN for detecting the tympanic membrane and training previously the CNN for the classification of tympanic membrane. The faster CNN is applied at first to the real images. The total images in the dataset are increased successively by expending a fundamental intensification technique like rotation and flipping.

The first and consequently removed tympanic film patches are at last info independently to the CNNs. Several already existing models are also available for the image data sets. These models might bring about a normal accuracy of 75.85% in the tympanic layer discovery. All CNNs in the characterization delivered palatable outcomes, with the projected methodology accomplishing a precision of 90.48% with the VGG-16 type model. This methodology could possibly been utilized in forthcoming ontological clinical choice emotionally supportive networks to expand the demonstrative precision of the doctors and decrease the general pace of misdiagnosis. Upcoming investigations will concentrate on expanding the quantity of tests in the eardrum data set to cover a full scope of ontological environments. This would empower to understand a multi-class grouping in OM diagnosis [50].

8.2.15   THORAX DISEASE CLASSIFICATION USING CNN

Here the assignment of thorax malady determination on chest X-beam (CXR) pictures is considered. Most existing techniques and large gain proficiency are observed in worldwide pictures as information. In any case, thorax illnesses typically occur in (little) confined zones which are malady explicit. In this way, preparing CNNs utilizing worldwide pictures might be influenced by the (exorbitant) unimportant loud territories. Moreover, because of the poor arrangement of some CXR pictures, the presence of sporadic fringes upsets the system execution. For tending to the above issues, a new scheme is proposed to coordinate the worldwide and neighborhood signs into a three-branch attention guided convolution neural networks (AG-CNN) to distinguish thorax infections. A consideration-guided veil surmising based editing technique is proposed to maintain a strategic distance from commotion and improve arrangement in the worldwide branch. AG-CNN likewise coordinates the worldwide signals to repay the lost discriminative prompts by the nearby office. In particular, proficiency is gained initially with a worldwide CNN branch utilizing worldwide pictures. At that point, guided by the consideration heat map produced from the worldwide branch, deduction a veil to trim a discriminative area from the worldwide picture. The neighborhood district is utilized for preparing a nearby CNN branch. At the end, connect the last pooling layers of both the worldwide and nearby offices for calibrating the combination branch. Investigations on the ChestX-ray14 dataset exhibit that in the wake of coordinating the nearby prompts with the worldwide data, the normal AUC scores are improved by AG-CNN [51].

8.2.16   BLOOD CELL CLASSIFICATION USING CNN

Classification of white blood cell image plays as a major category for the diagnosis of medical images. Here a CNNs scheme is proposed in order to classify WBC images. As like previous approaches, here also five different layers (1 convolutional, 3 max-pooling and 1 fully connected) are used [53]. Hyperspectral imaging is a rising imaging methodology in the field of clinical applications, and the mix of both otherworldly and spatial data gives riches data to cell characterization. Deep CNN is utilized to accomplish platelet segregation in clinical hyperspectral pictures (MHSI). As a profound learning design, CNNs are relied upon to get progressively discriminative and semantic highlights, which impact order precision to a limited degree. Test results dependent on two genuine clinical hyperspectral picture informational indexes, put forth that cell order utilizing CNNs is viable. Furthermore, contrasted with customary help vector machine (SVM), which together endeavors spatial and ghostly highlights, can accomplish better arrangement execution, displaying the CNN-based strategies' gigantic potential for exact clinical hyperspectral information order [54]. A clinical hyperspectral picture with high spatial and ghastly goals may contain more cell data; however, picture preparing may exist intricacy and impedance due to the high measurement. So as to utilize the data more proficient, the dimensionality decrease techniques to lessen the picture measurement, for example, the work of art PCA is utilized. PCA is acquainted with diminish repetitive unearthly data and lessen the information measurement to a satisfactory scale. Since the structured analysis utilizes the CNN, complete use of the way that neighboring pixels in MHSI will in general have a place similar class also, separate an enormous number of little districts (a focal pixel with its neighbors) [55]. Through the on-going advancement of profound learning, CNN has become serious and far superior execution than person in some picture handling issue and its ability motivates to contemplate the chance of applying CNN for hyperspectral picture arrangement utilizing the otherworldly marks. The CNN shifts in how the convolutional and max-pooling layers are acknowledged and how the nets are prepared.

8.3 CONCLUSION

The various approaches corresponding to the classification of the medical images is been reviewed, and an inclusive analysis is done with certain frameworks that includes CNNs. Here many outstanding works have been analyzed and revived. From this, it is observed that by applying CNN for the medical image classification yields some extraordinary results showing various disease patterns that makes a promising approach for the improvement of the diagnosis of various health issues. Some of the approaches mentioned here is thorax disease, tympanic membrane detection, brain tumor, liver lesion, AD, tomography images, breast cancer, breast cancer lesion, TB identification, abnormality detection in WCE images, skin lesions, sound sequence, DNA damage, and knee cartilage segmentation. This chapter has given the point by point audit of image classification systems for determination of human diseases by the classifying the medical images from the datasets by using the CNNs upsides and downsides for every procedure and also research perspective points are discussed which paves the way for future enhancement in medical image classification and disease identification.

KEYWORDS

- attention guided convolution neural networks
- contrast-enhanced computed tomography
- deep neural network
- image datasets
- machine learning
- medical image

REFERENCES

1. Li, Q., et al., (2014). Medical image classification with convolutional neural network. In: *2014 13th International Conference on Control Automation Robotics and Vision (ICARCV)*. IEEE.
2. Prasoon, A., et al., (2013). Deep feature learning for knee cartilage segmentation using a triplanar convolutional neural network. *International Conference on Medical Image Computing and Computer-Assisted Intervention*. Springer, Berlin, Heidelberg.
3. Raj, A., et al., (2018). Automatic knee cartilage segmentation using fully volumetric convolutional neural networks for evaluation of osteoarthritis. In: *2018 IEEE 15th International Symposium on Biomedical Imaging (ISBI)*. IEEE.
4. Heimann, T., et al., (2010). Segmentation of knee images: A grand challenge. *Proc. MICCAI Workshop on Medical Image Analysis for the Clinic*.
5. Stammberger, T., et al., (1999). Determination of 3D cartilage thickness data from MR imaging: Computational method and reproducibility in the living. *Magnetic Resonance in Medicine: An Official Journal of the International Society for Magnetic Resonance in Medicine, 41*(3), 529–536.

6. Ambellan, F., et al., (2019). Automated segmentation of knee bone and cartilage combining statistical shape knowledge and convolutional neural networks: Data from the osteoarthritis initiative. *Medical Image Analysis, 52*, 109–118.

7. Deniz, C. M., et al., (2018). Segmentation of the proximal femur from MR images using deep convolutional neural networks. *Scientific Reports, 8*(1), 1–14.

8. Einhorn, T. A., Joseph, A. B., & Regis, J. O., (2007). *Orthopaedic Basic Science: Foundations of Clinical Practice*. Amer Academy of Orthopedic.

9. Lakhani, P., & Baskaran, S., (2017). Deep learning at chest radiography: automated classification of pulmonary tuberculosis by using convolutional neural networks. *Radiology, 284*(2), 574–582.

10. Atila, Ü., et al., (2020). Classification of DNA damages on segmented comet assay images using convolutional neural network. *Computer Methods and Programs in Biomedicine, 186*, 105192.

11. Zhou, S. K., Hayit, G., & Dinggang, S., (2017). *Deep Learning for Medical Image Analysis*. Academic Press.

12. Anders, F., Mario, H., & Mirco, F., (2020). Automatic classification of infant vocalization sequences with convolutional neural networks. *Speech Communication*.

13. Khamparia, A., et al., (2019). Sound classification using convolutional neural network and tensor deep stacking network. *IEEE Access, 7*, 7717–7727.

14. Dey, N., et al., (2017). Wireless capsule gastrointestinal endoscopy: Direction-of-arrival estimation based localization survey. *IEEE Reviews in Biomedical Engineering, 10*, 2–11.

15. Sadasivan, V. S., & Chandra, S. S., (2019). High accuracy patch-level classification of wireless-capsule endoscopy images using a convolutional neural network. In: *2019 IEEE 16th International Symposium on Biomedical Imaging (ISBI)*. IEEE.

16. Krizhevsky, A., Ilya, S., & Geoffrey, E. H., (2012). ImageNet classification with deep convolutional neural networks. *Advances in Neural Information Processing Systems*.

17. Fan, X., et al., (2019). Effect of image noise on the classification of skin lesions using deep convolutional neural networks. *Tsinghua Science and Technology, 25*(3), 425–434.

18. Coudray, N., et al., (2018). Classification and mutation prediction from non-small cell lung cancer histopathology images using deep learning. *Nature Medicine, 24*(10), 1559–1567.

19. Mohammadian, S., Ali, K., & Yaser, M. R., (2017). Comparative study of fine-tuning of pre-trained convolutional neural networks for diabetic retinopathy screening. In: *2017 24th National and 2nd International Iranian Conference on Biomedical Engineering (ICBME)*. IEEE.

20. Abideen, Z. U., et al., (2020). Uncertainty assisted robust tuberculosis identification with Bayesian convolutional neural networks. *IEEE Access, 8*, 22812–22825.

21. Rajaraman, S., & Sameer, K. A., (2020). Modality-specific deep learning model ensembles toward improving TB detection in chest radiographs. *IEEE Access, 8*, 27318–27326.

22. Zabihollahy, F., Schieda, N., & Ukwatta, E., (2020). Patch-based convolutional neural network for differentiation of cyst from solid renal mass on contrast-enhanced computed tomography images. *IEEE Access, 8*, 8595–8602.

23. Khalifa, N. E. M., et al., (2020). Artificial intelligence technique for gene expression by tumor RNA-seq data: A novel optimized deep learning approach. *IEEE Access, 8*, 22874–22883.

24. Tzu-An, S., et al., (2019). Graph convolutional neural networks for Alzheimer's disease classification. In: *2019 IEEE 16th International Symposium on Biomedical Imaging (ISBI 2019)*. IEEE.

25. Petersen, R. C., et al., (2010). Alzheimer's disease neuroimaging initiative (ADNI). *Neurology, 74*(3), 201–209.

26. He, H., et al., (2008). ADASYN: Adaptive synthetic sampling approach for imbalanced learning. In: *2008 IEEE International Joint Conference on Neural Networks (IEEE World Congress on Computational Intelligence)*. IEEE.

27. Chawla, N. V., et al., (2002). SMOTE: Synthetic minority over-sampling technique. *Journal of Artificial Intelligence Research, 16*, 321–357.

28. Kesim, E., Zumray, D., & Tamer, O., (2019). X-ray chest image classification by a small-sized convolutional neural network. In: *2019 Scientific Meeting on Electrical-Electronics and Biomedical Engineering and Computer Science (EBBT)*. IEEE.

29. Gündüz, K., et al., (2018). Classification of tumor regions in histopathological images using convolutional neural networks. In: *2018 26th Signal Processing and Communications Applications Conference (SIU)*. IEEE.

30. Le, C. Y., et al., (1989). Handwritten digit recognition: Applications of neural network chips and automatic learning. *IEEE Communications Magazine, 27*(11), 41–46.

31. Hubel, D. H., & Torsten, N. W., (1959). Receptive fields of single neurones in the cat's striate cortex. *The Journal of Physiology, 148*(3), 574–591.

32. Sermanet, P., & Yann, L., (2011). Traffic sign recognition with multi-scale convolutional networks. In: *2011 International Joint Conference on Neural Networks*. IEEE.

33. Webb, W. R., Nestor, L. M., & David, P. N., (2014). *High-Resolution CT of the Lung*. Lippincott Williams & Wilkins.

34. Kermany, D. S., et al., (2018). Identifying medical diagnoses and treatable diseases by image-based deep learning. *Cell, 172*(5), 1122–1131.

35. Rawat, W., & Zenghui, W., (2017). Deep convolutional neural networks for image classification: A comprehensive review. *Neural Computation, 29*(9), 2352–2449.

36. Cardoso, M. J., et al., (2017). Intravascular imaging and computer-assisted stenting, and large-scale annotation of biomedical data and expert label synthesis. *CVII-STENT and Second International Workshop, LABELS*.

37. Deng, J., et al., (2009). ImageNet: A large-scale hierarchical image database. In: *2009 IEEE Conference on Computer Vision and Pattern Recognition*. IEEE.

38. Stanford Vision Lab, (2010). *ImageNet Summary and Statistics*. http://www.image-net.org/about-stats.

39. Jiang, T., et al., (2018). Data augmentation with Gabor filter in deep convolutional neural networks for SAR target recognition. *IGARSS 2018–2018 IEEE International Geoscience and Remote Sensing Symposium*. IEEE.

40. Yang, J., & Zhenming, P., (2013). SAR target recognition based on spectrum feature of optimal Gabor transform. In: *2013 International Conference on Communications, Circuits and Systems (ICCCAS)* (Vol. 2). IEEE.

41. Lai, Z. F., & Hui-Fang, D., (2018). Medical image classification based on deep features extracted by deep model and statistic feature fusion with multilayer perceptron□. *Computational Intelligence and Neuroscience, 2018*.

42. Beutel, J., et al., (2000). *Handbook of Medical Imaging: Medical Image Processing and Analysis, 2*. SPIE Press.

43. El Atlas, N., Mohammed, E. A., & Mohammed, W., (2014). Computer-aided breast cancer detection using mammograms: A review. In: *2014 Second World Conference on Complex Systems (WCCS)*. IEEE.

44. Ting, F. F., Yen, J. T., & Kok, S. S., (2019). Convolutional neural network improvement for breast cancer classification. *Expert Systems with Applications, 120*, 103–115.

45. Yang, Z., et al., (2019). EMS-net: Ensemble of multiscale convolutional neural networks for classification of breast cancer histology images. *Neurocomputing, 366*, 46–53.

46. Liu, M., et al., (2020). A multi-model deep convolutional neural network for automatic hippocampus segmentation and classification in Alzheimer's disease. *NeuroImage, 208*, 116459.

47. Roy, K., et al., (2019). Patch-based system for classification of breast histology images using deep learning. *Computerized Medical Imaging and Graphics 71*, 90–103.

48. Frid-Adar, M., et al., (2018). GAN-based synthetic medical image augmentation for increased CNN performance in liver lesion classification. *Neurocomputing, 321*, 321–331.

49. Amin, J., et al., (2020). Brain tumor classification based on DWT fusion of MRI sequences using convolutional neural network. *Pattern Recognition Letters, 129*, 115–122.

50. Zafer, C. B. E., & Yüksel, Ç., (2020). Convolutional neural network approach for automatic tympanic membrane detection and classification. *Biomedical Signal Processing and Control, 56*, 101734.

51. Guan, Q., et al., (2020). Thorax disease classification with attention guided convolutional neural network. *Pattern Recognition Letters, 131*, 38–45.

52. Zahedinasab, R., & Hadis, M., (2018). Using deep convolutional neural networks with adaptive activation functions for medical CT brain image classification. In: *2018 25th National and 3rd International Iranian Conference on Biomedical Engineering (ICBME)*. IEEE.

53. Banik, P. P., Rappy, S., & Ki-Doo, K., (2019). Fused convolutional neural network for white blood cell image classification. In: *2019 International Conference on Artificial Intelligence in Information and Communication (ICAIIC)*. IEEE.

54. Li, X., et al., (2017). Cell classification using convolutional neural networks in medical hyperspectral imagery. In: *2017 2nd International Conference on Image, Vision and Computing (ICIVC)*. IEEE.

55. Li, W., et al., (2016). Hyperspectral image classification using deep pixel-pair features. *IEEE Transactions on Geoscience and Remote Sensing, 55*(2), 844–853.

CHAPTER 9

MELANOMA DETECTION ON SKIN LESION IMAGES USING K-MEANS ALGORITHM AND SVM CLASSIFIER

M. JULIE THERESE,[1] A. DEVI,[2] and G. KAVYA[3]

[1]Assistant Professor, Department of ECE, Sri Manakula Vinayagar Engineering College, Pondicherry University, Puducherry, Tamil Nadu, India, E-mail: julietherese@smvec.ac.in

[2]Assistant Professor, Department of ECE, IFET College of Engineering, Villupuram, Tamil Nadu, India, E-mail: deviarumugam02@gmail.com

[3]Professor, Department of ECE, S. A. Engineering College, Chennai, Tamil Nadu, India, E-mail: gkavya2013@gmail.com

ABSTRACT

Nowadays, skin cancer has become common among people. Skin cancer is the uncontrolled growth of cells within the skin and they can even spread to the other parts of the body and become fatal. The three broad types of skin cancer are squamous cell carcinoma, basal cell carcinoma, and melanoma. Mortality due to melanoma are maximum among patients and hence melanoma which is kind of skin cancer is dangerous when it grows outside the epidermis. This seems to be threatening in comparison to squamous cell carcinoma, basal cell carcinoma. Skin cancer can be detected during the primitive stage by non-invasive computerized dermoscopy. The proposed system is to detect melanoma using image processing that involves certain procedure, first stage is preprocessing followed by segmentation, next is feature extraction, and finally classification. The algorithm proposed for each step is the Sobel process, Otsu's method, ABCD rule, and K-means with support vector machine (SVM) classifier. These algorithms, when implemented together, give good accuracy in terms of the values of size and shape, color, and texture of the lesion. This leads to extract the region of interest, which is then utilized for computerized surgery. PH2 dataset is used for producing the results.

9.1   INTRODUCTION

9.1.1   IMAGE PROCESSING

Images are considered as double valued, 2D functions, f (r, s), where r and s represents the spatial coordinate system, and in the coordinates (r, s) the amplitude off is said to be gray level or intensity of the image at that point. An image is called a digital image when the values of r, s, and amplitude values of f are finite and discrete quantities. A digital image processed by the digital computer is referred to as digital image processing. Each digital image possesses a definite number of elements, which has a specific location and value. These elements indicate picture elements, image elements, and pixels. In short, pixel is a component of digital images.

9.1.1.1   DIGITAL IMAGE REPRESENTATION

Virtual image, a point adjacent to the mirror, on its existence it would radiate rays on both side of the mirror.

1. **Vector Images:** An image can be characterized by numbers to declare its contents using geometric forms size and position and shapes like rectangles, circles, lines, etc. Such images are known as vector images.
2. **Co-Ordinate Systems:** The image coordinate system is description is required to understand the processing of an image. The elements in an image is located with relative to every other element as shown in Figure 9.1 in the coordinate system is this is said to be the userspace, as users makes use of these coordinate systems to describe position of elements to each other.

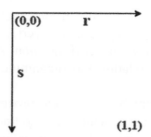

FIGURE 9.1 Coordinate system.

3. **Bitmap Images:** These are graphics stored in computer video memory. It is a form of digital photographs which symbolizes natural images along with various graphics. The phrase bitmap denotes the pattern of bits that forms pixel maps of a specified color. Images in bitmap are always represented in array form each element in the array is named as pixel element, that's corresponds to the color of a particular part of the image. The horizontal track in the image signifies a scan line.

4. **Raster Image:** The letter 'a' in the rasterized form is magnified 16 times using the pixel. As shown in Figure 9.2, the alphabet 'a' can be represented as a 12×14 matrix. The brightness of pixels (picture elements) is depicted using the values in the matrix. The larger values resemble the brighter areas, whereas darker areas represent lower values.

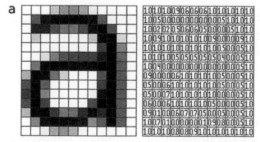

FIGURE 9.2 Raster image.

5. **Resolution:** The measurement of sampling density is termed as resolution. The resolution of bitmap images gives the relation between the physical dimension of the image and pixel dimension. It's measured by pixels per inch.

6. **Megapixels:** The entire pixels in the caught image are referred to as megapixels. The samples in the vertical and horizontal directions are known as the raster dimension. A megapixel image can be represented by relating the magnitude of the image with its dimensions.

7. **Scaling/Resampling:** Scaling is the process of creating different dimensions of an image. Scaling is also known as resampling. The reconstruction of the original continuous image and creating a new sample grid is done by resampling algorithms.

8. **Scaling Image Down:** The raster dimensions can be reduced by a process defined as decimation. It is carried out be done by taking an average for the values of the input pixels that contributes to each and every output pixel.

9. **Scaling Image Up:** The dimensions of the raster can be increased by a process called resolution enhancement. The process reproduces the original image.

10. **Color:** In computer graphics, the very usual way to design a RBG color model, is to resemble both LCD screens/projectors and CRT monitors reproduce color. The amount of red, green, and blue represent three values of each pixel. The RBG color image uses three times as much as the memory of a gray-scale of the similar pixel dimensions.

9.1.1.2   IMAGE FORMATION

The discrete representation of data processing of the spatial and intensity information is denoted as a digital image. The image here is considered to be treated as multidimensional signals. The steps involved in image formation are:

1. The very small number of essential elements summarizes the image formation process.

2. It is formalized in terms of functional representation mainly of a mathematical model.

3. The functional representation includes the object function (o), the point spread function (PSF) and in addition, the image will contain the additive noise (n).

4. **Object Function (o):** This function describes the capturing process by which the object is being imaged. It also determines the path of the light intensity of the object in the imaging instrument.

5. **The Point Spread Function (PSF):** It is the deterministic function that obtains the characteristics of the imaging instrument. As the output of the data is recorded, the route where the information of the target function exists is PSF.

6. **Noise (n):** A non-deterministic signal that has stochastic function is an additive noise, this is described as Gaussian noise distribution because of its statistical distribution. Noise (n) occurs in the recording phase of the image; all the unwanted noise distribution that occurs will form the consequences of additive noise.

9.1.1.3 FUNDAMENTAL STEPS INVOLVED IN DIGITAL IMAGE PROCESSING

1. **Image Acquisition:** It is the initial stage towards image processing, where the image is captured by the camera converts the optical information to digital data for further processing by a computer.

2. **Image Enhancement:** The easiest and very interesting part of digital image processing is image enhancement. The foremost idea of this technique is to carry out the part that is concealed or merely to stress on some characteristics of importance in an image.

3. **Image Restoration:** It is the process where an image's appearance can be enriched, and the technique is called image restoration. However, image restoration is said to be objective and image enhancement is said to be subjective. This technique is based on mathematical models like probability of image degradation.

4. **Color Image Processing:** It is an area that is increasingly gaining name attributable to the main increasingly usage of digital pictures over the internet.

5. **Wavelets:** An image represented by different degrees of resolution.

6. **Compression:** The technique used to reduce the storage mandatory to keep an image or otherwise a bandwidth to conduct compression.

7. **Morphological Processing:** The extraction of image modules which are suitable in representation, depiction, and shape.

8. **Segmentation:** The process of partitioning an image into smaller component objects or parts. Autonomous segmentation is the very tedious tasks in digital image processing.

9. **Representation and Depiction:** It is the next stage of segmentation, which is sensitive pixel data, constituting almost all the points in the area of interest or boundary of a region.

10. **Recognition:** Based on descriptors, the method that allocates a label to an object is known as recognition.

9.1.2 THE COMPONENTS OF AN IMAGE PROCESSING SYSTEM

1. **Sensing:** In accordance with sensing, two features that are obligatory to obtain digital images are a physical device and a digitizer. While capturing an image, a device that is physically available will be delicate and sensitive because of the energy emitted by the capturing entity, and hence the physical sensing device's output is converted onto digital data by the digitizer.

2. **Specialized Image Processing Hardware:** A specified hardware for image processing is composed of a digitizer. This hardware is also denoted as a front-end module, where they perform only primitive operations like arithmetic logic unit (ALU). The distinguishing characteristics of this system are speed that is the averaging process of images is as quickly as digitizing.

3. **Computer:** The general-purpose computers that range from PC to supercomputers are being used. Also in offline mode image processing responsibilities, any kind of well-resourced PC type is efficient to perform. In certain applications, particularly specialized designed computers are required to obtain high-level performance.

4. **Software:** The specialized modules that perform a specific task are the software used for image processing. A well-designed package utilizes the specialized modules, which include the ability to write a code by the user. More sophisticated software packages permit the integration of modules and general-purpose software commands from any one of the computer languages.

5. **Hardcopy:** The recording of images comprises of laser printers, inkjet units, film cameras, and digital terminals which organize a CD-Read Only Memory and optical disks as hardcopy devices.

6. **Image Displays:** In recent days, the use of image displays is mainly color. The integral parts of the computer system are image output and graphic card display which plays a vital role in screen image display.

7. **Networking:** The default function in any computer system is networking. Bandwidth is considered a key in image broadcast because in image processing applications, an enormous amount of data is essential.

8. **Storage:** The storage capability is the key application in image processing. Storage can be measured in highest to lowest units terabytes, gigabytes, megabytes, kilobytes, and bytes. The memory of the computer provides short-term storage. The frame buffers known as specialized boards, which store numerous images at video rates can rapidly be accessed. The final method includes a virtual image scroll, pan, and zoom. The principal categories for numerical storage in image processing are:

   • When the image is processed only short term memory is used;
   • For fast recall online storage is essential;
   • Infrequent access is categorized by archival storage.

In dealing with thousands and millions of images, providing adequate storage in an image processing system is considered to be a challenge because it deals with millions and billions of images providing suitable storage.

9.1.3   IMAGE PROCESSING TECHNIQUES

The image processing techniques includes the following subsections.

9.1.3.1   ACQUISITION

   • The physical quantities describing the internal aspects of the body or object which is created as a digital image is referred to as image acquisition.
   • It is considered as the first and foremost step of image processing because it contains the original information of the image that is captured.
   • It can be done by using a single sensor, sensor strips, and sensor array.

9.1.3.1.1   *Methods of Acquisition*

i.   **Laser Ranging Systems:** The principle of laser ranging system is that, towards the receiver side, the laser light will be reflected from the surface of the object. They calculate the depth by measuring the time and phase difference between the transmitter and receiver.

ii. **Moire Fringe Methods:** Very accurate depth data can be produced by Moire fridge methods. The flow of this method is that the grating is projected onto the object and over some reference grating, the image is formed in the plane. Then, by interfering with the image with reference grating, there will be a formation of Moire fringe contour pattern which is composed of dark and light stripes. By analyzing this pattern, any changes in the shape and depth can be identified.

iii. **Shape from Shading Methods:** This method is also used for calculating depth. The photometric stereo techniques are employed here in order to study the change in brightness of the object when it is placed in a fixed position but in different lighting environments and so that the certain depth measurements can be done. This method is not efficient while considering three-dimensional depth data acquisition.

iv. **Passive Stereoscopic Methods:** The technique employed in this method is stereoscopy. It measures the range from triangulation to the location already selected by two cameras for an imagined scene. The drawback of this method is to find the similarity level in any two images.

v.   **Active Stereoscopic Methods:** This method is proposed to overcome the limitations of the passive stereoscopic method. So here both the cameras can observe the scene illumination with a good source of light. The laser light source is also employed in this method.

9.1.3.2   *SEGMENTATION*

- The process where the image is divided into their constituent sub-regions is referred to as segmentation.
- The segmentation can be done by manual and automation, the manual segmentation is not as efficient as automation as it is a time-consuming task.

- The three main problems in the segmentation of medical images include noise, intensity non-uniformity, and partial volume averaging, and hence these problems are prevailed by the methods of segmentation.

9.1.3.2.1 Types of Segmentation

1. **Supervised Segmentation:** In this segmentation, the operator interaction is required throughout the completion of the process.
2. **Unsupervised Segmentation:** Whereas in unsupervised segmentation, the operator interaction is required only after the completion of the process, which is preferred over other techniques in order to provide reproducible results.

9.1.3.2.2 Methods for Segmentation

The methods of segmentation in medical images include three-generation which adds a certain level of algorithmic complexity in each generation:

1. **First Generation:** The analysis of image composed of low-level techniques and simplest forms come under the first generation. The three main segmentation problems which are mentioned above are subjected to this generation:
    i.   Thresholds are applied in order to distinguish regions in an image with intensity levels.
    ii. Region growing allows us to define the volume of the object and grows the region by identifying the similar pixels.
    iii. Edge tracing can be done by representing an object boundary by edge detection by means of forming an edge image with edge pixels and adjacent neighbor connectivity.
2. **Second Generation:** The slight modification in the application of optimization methods and uncertainty models, which includes efforts to overcome the problems in segmentation is the second-generation method.
    i.   Statistical pattern recognition is a mixture model where every pixel of the image is modeled as belonging to a single known set of classes.

ii.   C-means clustering identifies the features and groups the image pixels together. The feature will be represented between 0's and 1's, or else the cluster in numbers assumed is to be known for any medical images.

iii.   Deformable models within an image will be artificial, contrast over duration or closed surface which can be expanded, and hence they conform to certain features.

3.   **Third Generation:** The methods that incorporate higher knowledge levels are the Third generation model. This model produces automatic and accurate segmentation. The active shape model (ASM), active accurate model (AAM), shape model, rule-based segmentation, atlas-based segmentation, coupled surfaces is the updates in this generation.

9.1.3.3   FEATURE EXTRACTION

Based on the type of features, the classification and feature extraction includes the three main approaches are: statistical approach, spectral approach and structural or syntactic approach:

- The set of statistically extracted features in multidimensional feature space represented as a vector is referred to as a statistical approach.
- For processing random patterns, the statistical approach methods are particularly useful.
- The placement rules which are spatially organized to generate a complete pattern by texture primitives are referred to as a syntactic approach.
- In the syntactic approach, between the syntax of the language and structural pattern, a formal analogy is drawn.
- The better result will be provided by the structural approach while considering complex patterns.
- When the textures are evaluated by autocorrelation function and are defined by spatial frequency, then it is referred to as a spectral approach.
- In comparison with the other two approaches, a spectral approach is less efficient in any kind of pattern.
- The common features extracted from any biological images are the length to width ratio, standard deviation, mean, Eccentricity, Entropy, relative areas, and relative length.

9.1.3.4 CLASSIFICATION

- In general, the basic classifications of image processing are supervised and unsupervised classifications.
- In supervised classification, the classes are identified by the analyst and the computer is directed to classify them accordingly.
- In unsupervised classification, without the input of the analyst, the classes are made by considering the digital numbers or the properties of the spectrum.
- The medical image classification has an important role in teaching and diagnostic purpose in medicine, which is the key technique of computer-aided diagnosis (CAD) where the medical images are classified based on the texture in various regions of the image.

9.1.4 THE MATHEMATICS OF IMAGE FORMATION

The process of image formation can be characterized mathematically. Mathematically, the arrangement of a picture is a procedure that changes an input distribution into an output distribution as shown in Figure 9.3. The transformation of the spatial distribution of light in one domain (object plain) to distribution in another plane (image plane) can be viewed as a system of a simple lens.

FIGURE 9.3 Mathematical model for image processing.

Any imaging gadget is a framework, or a "black box," whose properties are characterized by the way in which an input distribution is mapped to an output distribution. The procedure by which an imaging framework changes the input to output can be seen from an alternative point of view, to be specific that of the Fourier or frequency domain.

9.1.5   DIGITAL IMAGE PROCESSING IN FIELD

The use of imaging is mainly based on gamma rays, which include astronomical observations and nuclear medicine. In nuclear medicine, the patient is injected with a radioactive isotope when decay gamma rays are emitted. One of the most commonly used sources of EM radiation for imaging is X-rays. It is specially used for medical diagnostics, but also used widely used in industry, astronomy, etc. The X-rays used for industrial and medical imaging are generated using an X-ray tube. Angiography one of the major applications in a zone called contrast enhancement radiography. This method is used to acquire images (called angiograms) of blood vessels. For example, a catheter is a small flexible hollow tube inserted into a vein or artery in the groin. The applications of ultraviolet (UV) "light" are diverse. They include biological observation, lasers, astronomical observations, and lithography. Similarly, there are fields such as computer vision to match the human vision, which includes learning, make inferences, and visual inputs based on actions. This is a branch of AI (artificial intelligence) which aims to emulate human intelligence.

9.1.6   ARTIFICIAL INTELLIGENCE (AI)

Artificial intelligence (AI) is the captivating and widespread field of computer science that has utmost opportunity in the future. This also has an affinity to make a machine to work has a human. At present AI works on a variety of subfields. Now-a-days, many of the technologies and gadgets that support those technologies works on AI. Best examples are self-driving car, streaming online music or movies and most of the social networks like Facebook, Orkut, Instagram makes use of artificial network.

The word *artificial* defines man-made and the word *intelligence* defines thinking power. AI could be stated in simple terms as man-made thinking power. It could even be defined as "a branch of computer science wherein one can create an intelligent machine that has thinking power and behaves like a human." With AI pre-programming a system would not be required. Yet a machine can be created with algorithms that can work in its own intelligence.

With AI, systems that use software that can solve real-world problems can be built. The Google Assistant, Siri, Alexa are some of the AI personal assistants. Robots can be designed with AI that could work as effectively as human's work.

➢ **Goals of AI:**
- Reproduce human intelligence.
- Capable of solving rigorous tasks.
- Capable of observing, sensing, and putting it into action like human's do.

Intelligence is composed of reasoning, learning, perception, problem solving, and linguistic intelligence.

➢ **Agents in AI:** The agents sense the information and reacts to the environment. They utilize sensors and actuators to perform tasks. An intelligent agent should possess properties like knowledge, belief, intension, etc. The agents could be Human-agent, Software agent, and Robotic agent.

➢ **Types Agents in AI:** Based on the degree of the perceived intelligence and capacities, the agents can be grouped into five classes: Simple Reflex, Model-Based, Goal-Based, Utility-Based, and Learning Agent. The simple reflex agent operates only on the present perception and is not dependant on the previous state of the system. Model-based agent operates only by discovering a rule whose condition matches with the present situation. The goal-based agent is flexible and makes decisions only based on the goal set. In addition to the goal-based agent, the utility-based agent provides parameters to measure the effective way to achieve the goal set by a goal-based agent. Unlike Simple reflex, the learning agent possesses learning capabilities and learns from the past experiences. There are five problem-solving steps that are expected to be handled by an agent they are, defining the problem, analyzing the problem, identifying suitable solutions, choosing the right solution, and finally, implementation.

➢ **Types of Search Algorithm:** Based on the search problems, search algorithm can be classified as: Uninformed search (Blind Search) and Informed search (Heuristic search). The uninformed search will never possess the domain knowledge and works in a brute force way. Informed search always possess the domain knowledge and proves to be the most efficient method in terms of providing a good solution in reasonable time, but the obtained solution might not always be the best solution.

9.1.7 MACHINE LEARNING

Machine learning is a subgroup of AI in which the created algorithms can modify itself without much intervention of human to produce the desired output by following a structured data. For a machine to learn the data that has to be fed should be labeled and differentiated based on specific features, the machine learning algorithm learns the same and works on the understanding of the labeled data, thereby classifying the required output. Learning strategy here could be supervised learning, unsupervised learning, or reinforcement learning. Supervised learning requires labeled data to understand the feature to produce the output. Unsupervised learning is learning with unlabeled data. The next learning methodology is reinforcement learning that learns based on feedback, as shown in Figure 9.4.

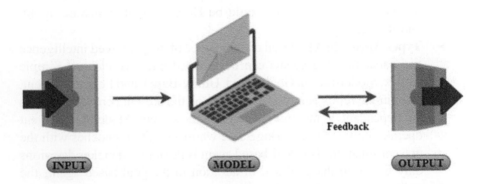

Feedback

INPUT   MODEL   OUTPUT

FIGURE 9.4   Process of machine learning.

In general, machine learning utilizes mathematical models and algorithms to discover and examine data to make estimates and take decisions accordingly; this enables the computers and humans to interact autonomously. Alan Turing, a pioneering computer scientist, published an article in 1950 based on *can machines think?* With this idea, the turing test was conducted in 1957, when Frank Rosenblatt designed the first neural network for computers. Following this, two neural network models were created by Bernard Boudreau and Marcin Hawke that could detect binary patterns and another to eliminate echo on phone lines. In 1967, the nearest neighbor algorithm was developed for pattern recognition. In the year 1981, Jarrell De Jong designed an algorithm wherein a computer analyzes a data and differentiates unimportant data. During the 1990's machine learning became a data-driven approach, where huge data were analyzed from which conclusions can be made.

9.1.7.1   MACHINE LEARNING ALGORITHMS

To interact with a computer to define the process it should do is done by writing a program. Programs are logics that are built using syntax, the logics are the algorithms. These algorithms are a set of rules or processes that are used for manipulation, especially in problem solving operations, based on the problem a particular algorithm is to be used. Algorithms could be anomaly detection algorithms, classification algorithms, clustering algorithms, regression algorithms, and reinforcement learning.

Classification algorithms are utilized for classifying a record that requires a reduced number of answers. Anomaly algorithms are used to analyze patterns and find the anomaly that breaks the pattern. If the resultant of a process needs to be a numeric value, then the regression algorithm is the best choice. Moving ahead with clustering algorithms, these algorithms are utilized to interpret the assembly of a dataset, where the data is separated into clusters that facilitate easy understanding of structured dataset. The next algorithm is the reinforcement algorithm that acts similarly to a human brain; the algorithm produces output based on the past inputs (pervious history) and present input. They gain learning from these outputs and make decisions for the next action.

9.1.8   PROPOSED MODEL BLOCK DIAGRAM

Figure 9.5 shows the proposed model that uses PH2 dataset for analysis, Sobel process for Edge detection, Otsu's method for segmentation, ABCD Rule for feature extraction, K-means, and SVM classifier.

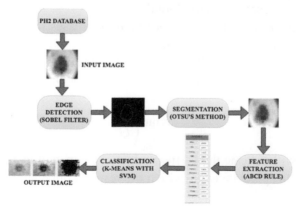

FIGURE 9.5   Proposed model.

9.1.8.1 PH2 DATASET

Data is raw information; it is the representation of human and machine observations. In this chapter, since the discussion is about Melanoma, PH2 database (DB) has been used; this is a dermoscopic image DB that is being used for CAD. The lesion images have pixels of 768×560 resolutions and the dataset has totally 200 images of melanoma. Datasets are stored separately based on different parameters like color segmentation, Lesion segmentation, pigment network, regression areas, etc. This DB allows for processing like segmentation and classification of images.

9.1.8.2 EDGE DETECTION

In image processing, the Sobel Feldman operator or Sobel filter is used for the purpose of detecting the edges of the image, where an image emphasis edges are created. The main function of the Sobel filter is, it calculates the gradient of image intensity within the image at each and every pixel. The gradient magnitude of the image is found by combining the two gradients G_x and G_y at each pixel and hence G is:

$$G = \sqrt{G_x^2 + G_y^2}$$

where; G_x and G_y will be the vertical and horizontal derivatives of the gradient at each pixel of the image.

G_x = {[3×3] kernel} * A-Vertical.
G_y = {[3×3] kernel} * A-Horizontal.

where; A-original input image; x coordinate-increases in right side; and y coordinate-increases in the downwards.

By computing G_x and G_y, the gradient of image intensity can be calculated in the output image. The direction of gradient is calculated by:

$$\theta = \arctan\left(\frac{G_y}{G_x}\right)$$

where; $\theta = 0$ will indicate the vertical edge. The constant intensity will be a zero vector to the pixel on edge, which the result is obtained.

9.1.8.3 SEGMENTATION

Otsu technique is considered to be successful and is one of the effective methods of segmentation because of its thresholding way for binarization in the processing of an image. It performs automatic thresholding as it returns a single intensity threshold. It finds the optimal threshold value of the image by going through all the possible values from 0 to 255. The thresholding in image processing is done in order to split the images into segments for defining their respective boundaries. The pixels in the intensity threshold are divided into two classes:

1. Foreground; and
2. Background.

Of the above two classes, the threshold is defined as the weighted sum of variances that minimizes the intra-class variances. The histogram will be created for the input image initially, and then the variance will be found. Two-dimensional Otsu's method performs better in dealing with noisy images for calculation of threshold.

> **Otsu's Method Steps:**
> - **Step 1:** The intensity level of both histogram and probability are computed.
> - **Step 2:** Initially, $\mu_i(0)$ and $\mu_i(0)$ are set up.
> - **Step 3:** Step over all probable threshold $t = 1...$ maximum intensity:
>   - ω_i and μ_i are updated.
>   - $\sigma_c^2(t)$ is calculated.
> - **Step 4:** The maximum $\sigma_c^2(t)$ corresponds to the desire threshold.
> - **Step 5:** To maxima can be computed. The greater max is $\sigma_{c1}^2(t)$ and the greater or equal max is $\sigma_{c2}^2(t)$.
> - **Step 6:** The desired threshold is obtained by taking the average for the threshold value.

9.1.8.4 FEATURE EXTRACTION

The asymmetric, border structure, color difference, and diameter of the lesion (ABCD) are the basis for diagnosing of skin diseases. In image processing, feature extraction plays an important role in analysis and exploring the image, which depends on the ABCD rule of dermatoscopy. As it was invented by Stolz's et al., this rule is called a STOLZs method, which is mainly used for dermoscopic differentiation that is the field of detecting and treating skin cancer. The features of ABCD rules are as follows:

1. **Asymmetric Feature:** In image analysis, symmetry is the key feature which helps in determining the full image rid of the noise. The asymmetric value can be of two types: asymmetry index and the lengthening index. Asymmetry index is used in checking the degree of symmetry and hence AI is given as:

$$AI = \frac{1}{2}\sum_{h=1}^{2} \Delta Ah / A_L$$

Lengthening index is the factor for defining the elongation of a lesion and hence it is given as:

$$A = \lambda' / \lambda''$$

2. **Border Structure:** By calculating edge abruptness, fractal dimension, and compact index, the border structure can be analyzed.
    EDGE ABRUPTNESS is calculated by estimating the irregular boundaries and hence by finding the radial and mean distance the Cr is given as:

$$C_r = \frac{\frac{1}{P_L}\sum p \in C(d_2(p, H_L) - s_d)^2}{s_d^2}$$

Fractal dimension can be calculated by using the box-counting method where the images are divided into boxes. The value must be an integer, and for lines, files, and cubes, the values are 1, 2, and 3. Compactness index measures the barriers in 2D objects which are quite difficult and hence CI is given as:

$$CI = \frac{S_L^2}{4\pi A_L}$$

3. **Color Variation:** In the diagnosis of any kind of skin cancer or diseases, the color variation is considered as one of the important factors. The color variation is described as the color homogeneity

and the correlation between the geometry and photometry CPG. In order to evaluate the color variation in the lesion, the value of C_{pg} is essential which is given as:

$$C_{pg} = \frac{1}{M_L} \cdot \sum_{j \in L} \frac{(lum(j) - m_l).(d_2(j.G_L) - m_d)}{u_l.u_d}$$

4. **Diameter:** In order to calculate the diameter, through the midpoint, the pixel edges to the edge pixel should be drawn and then averaged. The final step of ABCD rule is to find the total dermatoscopic values (TDV) and the formula for calculating TDV is:

$$TDV = P \cdot 1.3 + Q \cdot 0.1 + R \cdot 0.5 + S \cdot 0.5$$

If TDV is 1.00 to 4.75 –Beginning skin lesion; If TDV is 4.75 to 5.45 –Suspicious; If TDV is >5.45 –Melanoma/possibilities to other skin diseases.

9.1.8.5 CLASSIFICATION

9.1.8.5.1 K-Means Algorithm

The classification of an image into dissimilar clusters is known as image segmentation. Clustering is defined as a group of data separated into a specific number of sets. Many studies have been carried out in the field of image segmentation by means of clustering or grouping. There are diverse methods, but the most standard method is the k-means clustering algorithm. It is an unsupervised algorithm used to segment the interest region from the background. A gathered data is partitioned into groups of k number of data by K-means clustering. The given set of data is classified into k number of disjoint clusters. This algorithm consists of two distinct phases. The centroid is calculated in the first phase, and each point is taken to the group or cluster that has the adjoining centroid from the data point individually. There are various approaches to describe the nearest centroid distance, but generally, the used technique is Euclidean distance. The new centroid for the individual clusters is recalculated once the grouping is done. Between each center and each data point, a new Euclidean distance is calculated further the points are assigned to which as minimum Euclidean distance. In partition, each cluster is defined by its member objects and centroid. The point at which the totality of distance from all objects is diminished. Hence K-means is also called an iterative algorithm.

Steps involved in K-means algorithm:

Let $S = \{S_1, S_2, S_3, S.........Sn\}$ be the set of data points $U = \{u_1, u_2, u_3.......u_d\}$ be the centers

- **Step 1:** The cluster centers d is selected randomly.
- **Step 2:** The distance between the cluster center and data points are calculated.
- **Step 3:** To the cluster center data points are assigned whose distance is minimum of all data center from the cluster center.
- **Step 4:** New cluster is recalculated:

$$u_i = (1/d_i)\sum_{j=1}^{d_i} S_i$$

where; d_i is the number of total data points in the ith cluster.

- **Step 5:** To obtain a new cluster, the distance between each data point is calculated.
- **Step 6:** The process is declined if no data point is reassigned, else Step 3 continues.

9.1.8.5.2   *Support Vector Machines (SVM)*

SVM are efficiently used as a classification tool in a wide range of applications. It is a binary classification algorithm that deals with the manipulation of images and also that it falls under the machine learning technique. The efficiency of SVM is proven over RBF classifiers and neural networks. The SVM operates on two mathematical operations, they are:

a. Non-linear mapping of an input vector that is hidden from both of the input and output into a high dimensional feature space.

b. The optimum hyperplane is constructed for the separation of features in Step 1. The equations used to determine the support vector is 1–8 [10, 11].

➢ **Variable Definition:**
- Considering the input space, let assume that the vector drawn is denoted by u and the dimension is given as m_0.

- Let $\{\varphi_k(U)\}$ denotes the non-linear transformation sets to the feature space from the input for $k=1$ to m_1.
- Let the dimension of feature space be m_1.
- Let $\{w_k\}$ be the linear set of weights that connects the feature space to the output space.
- Let $\{\varphi_k(U)\}$ denotes the input supplied via the feature space to the weight w1.
- Let b be the bias.
- α_1 is denoted as the Lagrange coefficient.
- d_i represents the target output.

➢ **Procedure Involved in the Design of SVM:**

- **Step 1:** At the decision surface, the hyper-plane is defined as:

$$\sum_{i=1}^{N} \alpha_i d_i K(u, u_i) = 0 \tag{1}$$

where; u_i: input pattern of ith class; u: Input vector; $K(u, u_i)$: the inner product of two vectors.

$$w = \sum_{i=1}^{N} \alpha_i d_i \varphi(u_i) \tag{2}$$

$$\varphi(u) = [\varphi_0(U), \varphi_1(U), \ldots\ldots \varphi_{m_i}(U)]^T \tag{3}$$

where; $\varphi_0(u) = 1$ for all u; W_0 represents the bias b.

- **Step 2:** The kernel function is chosen as a polynomial learning machine. In order to satisfy the Mercer's theorem, the kernel K (u, u_i) is required.

$$K(u, u_i) = (1 + u^T u_i)^2 \tag{4}$$

- **Step 3:** The Lagrange multiplier that minimizes the object function for $i = 1$ to N is required.

$$Q(\alpha) \sum_{i=1}^{N} = \alpha_i - \frac{1}{2} \sum_{i=2}^{N} \sum_{j=1}^{N} \alpha_i \alpha_j d_i d_j K(u_i, u_j) \tag{5}$$

Subject to the constraints:

$$\sum_{i=1}^{N} = \alpha_i d_i = 0 \tag{6}$$

$$0 \leq \alpha_i \leq C \tag{7}$$

for i = 1, 2, …… N.

- **Step 4:** In correspondence to the optimum value, the linear weight vector w_0 of the Lagrange multipliers are obtained using the formula.

$$w_0 \sum_{i=1}^{N} \alpha_{0,i} d_i \varphi(u_i) \tag{8}$$

where; $\varphi(u_i)$ is the feature space induced due to u_i. The optimum bias b_0 is represented by w_0.

9.1.9   RESULTS AND DISCUSSION

➤ **Step 1: Input Image:** Database input image is taken from the PH2 database, as shown in Figure 9.6.

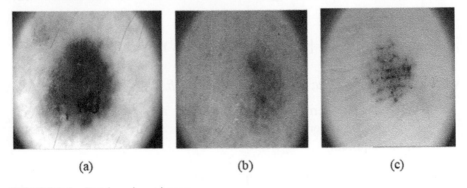

(a)   (b)   (c)

FIGURE 9.6   Database input image.

➤ **Step 2: Edge Detection:** Affected skin lesions are cropped to the depth region of the skin lesion that is shown in blue color and maximum and minimum intensity value beside the skin lesion is displayed as shown in Figure 9.7.

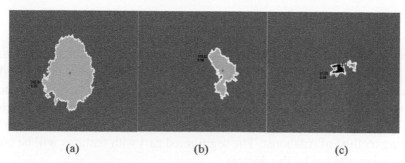

FIGURE 9.7   Edge detection.

➢ **Step 3: Segmentation:** The mask region is obtained using Otsu's method that particular masked region is called a segmented image, as shown in Figure 9.8.

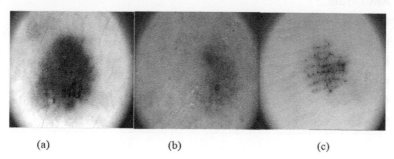

FIGURE 9.8   Segmented image.

➢ **Step 4: Feature Extraction and Classification:** Figure 9.9 depicts the output of segmented and features extraction values using K means integrated SVM and ABCD rule.

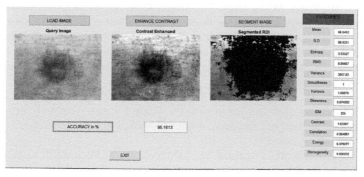

FIGURE 9.9   Feature extraction and classification.

9.2   CONCLUSION

With the growing rate of melanoma, the detection of skin lesions using computer-aided techniques to identify the patient with melanoma becomes necessary in the field of medicine. By the deployment of the Sobel process, Otsu's method, ABCD rule, K-means, and SVM classifier, the segmentation and classification result in an accuracy of 92% when compared to the existing method of Adaboost. The segmented part with features will be taken for decision making by the doctors.

KEYWORDS

- ABCD rule
- active accurate model
- image processing
- K-means
- melanoma
- Otsu's method
- Sobel process

REFERENCES

1. Julie, T. M., & Christo, A., (2020). A survey on melanoma: Skin cancer through computerized diagnosis. *International Journal of Advanced Research in Innovative Discoveries in Engineering and Applications (IJARIDEA), 5*(1), 9–18. ISSN(Online): 2456-8805.
2. Teresa, M., Pedro, M. F., Jorge, M., Andre, R. S. M., & Jorge, R., (2013). PH²-A dermoscopic image database for research and benchmarking. In: *35*[th] *International Conference of the IEEE Engineering in Medicine and Biology Society*. Osaka, Japan.
3. Satheesha, T. Y., Satyanarayana, D., Giri, P. M. N., & Kashyap, D. D., (2016). *Melanoma is Skin Deep: A 3D Reconstruction Technique for Computerized Dermoscopic Skin Lesion Classification*, 1–7. IEEE.
4. Abuzaghleh, O., Faezipour, M., & Barkana, B. D., (2015). A comparison of feature sets for an automated skin lesion analysis system for melanoma early detection and prevention. In: *Systems, Applications and Technology Conference (LISAT)* (pp. 1–6). 2015 IEEE Long Island.
5. Barata, C., Ruela, M., Francisco, M., Mendonça, T., & Marques, J. S., (2014). Two systems for the detection of melanomas in dermoscopy images using texture and color features. *Systems Journal, IEEE, 8*(3), 965, 979.

6. Glaister, J., Wong, A., & Clausi, D. A., (2014). Segmentation of skin lesions from digital images using joint statistical texture distinctiveness. *Biomedical Engineering, IEEE Transactions, 61*(4), 1220, 1230.

7. Sadeghi, M., Lee, T. K., McLean, D., Lui, H., & Atkins, M. S., (2013). Detection and analysis of irregular streaks in dermoscopic images of skin lesions. *Medical Imaging, IEEE Transactions, 32*(5), 849, 861.

8. Tittmann, B. R., Miyasaka, C., Maeva, E., & Shum, D., (2013). Fine mapping of tissue properties on excised samples of melanoma and skin without the need for histological staining. *Ultrasonics, Ferroelectrics, and Frequency Control, IEEE Transactions, 60*(2), 320, 331.

9. Zhuo, S., & Sim, T., (2011). Defocus map estimation from a single image. *Pattern Recognition, 44*(9), 1852–1858.

10. Maglogiannis, I., & Doukas, C. N., (2009). Overview of advanced computer vision systems for skin lesions characterization. *Information Technology in Biomedicine, IEEE Transactions, 13*(5), 721, 733.

11. Menzies, S., Ingvar, C., Crotty, K., & McCarthy, W., (1996). Frequency and morphologic characteristics of invasive melanomas lacking specific surface microscopic features. *Arch. Dermatol., 132*(10), 1178–1182.

12. Stolz, W., Riemann, A., & Cognetta, A., (1994). ABCD rule of dermatoscopy: A new practical method for early recognition of malignant melanoma. *Eur. J. Dermatol., 4*, 521–527.

13. Pehamberger, A. S., & Wolff, K., (1987). *In vivo* epi luminescence microscopy of pigmented skin lesions-I: Pattern analysis of pigmented skin lesions. *J. Amer. Acad. Dermatol., 17*, 571–583.

CHAPTER 10

ROLE OF DEEP LEARNING TECHNIQUES IN DETECTING SKIN CANCER: A REVIEW

S. M. JAISAKTHI[1] and B. DEVIKIRUBHA[2]

[1]Associate Professor, School of Computer Science and Engineering, Vellore Institute of Technology, Tamil Nadu, India, E-mail: jaisakthi.murugaiyan@vit.ac.in

[2]School of Computer Science and Engineering, Vellore Institute of Technology, Tamil Nadu, India, E-mail: devikiruba.b2018@vitstudent.ac.in

ABSTRACT

Automatic detection of skin cancer in the dermoscopic image is required for detecting melanoma at an early stage. Skin cancer is extremely dangerous, and it is curable when diagnosed at an early stage. The most common cause of skin lesions is due to the infection of the skin by bacteria, viruses, fungi, or parasites. Previously, skin cancer was diagnosed by biopsy. Later the lesions occurring on the surface of the skin can be detected by analyzing the dermoscopic images of the skin. However, doing manual analysis is time consuming, and it is also difficult to differentiate benign from malignant. Any automated skin detection system can aid in the early detection and also can help in differentiating the tumors. However, attaining high sensitivity and specificity is still challenging due to the unavailability of more malignant samples. In order to cure the human skin cancer with accuracy and also to reduce the error-prone due to human, dermatoscope is in use nowadays. In this review, we discussed the state-of-the-art deep learning (DL) form the foundation of machine learning as it provides better accuracy for medical images.

10.1 INTRODUCTION

Human skin is the largest integumentary system which gives outer covering. Cancer, a malignant tumor develops in the skin due to abnormal cell present in the epidermis. Nearly 290,000 malignant melanoma cancers [1] has been diagnosed and treated globally in 2018. Moreover, 11000 cases are treated, but it is twice more than the past 10 years and will be doubled by 2045 as in [3]. Identifying and initiating treatment in the early-stage would have a positive impact [2] to prevent the melanoma from further infection. Major problem in skin cancer is exposure to sunlight [3], i.e., ultraviolet (UV) radiations. It is agreed that suntan makes human healthy. However, one of the three people is affected [3] due to the sun. Some of the changes in the skin are bleeding, open sore, pink/red translucent bump, and rough surface.

Clinically, melanoma can be of four types [4] as: (i) superficially melanoma; (ii) lentigo maligna melanoma; (iii) nodular melanoma; and (iv) acral lentiginous melanoma. We can also find the melanoma in the parts of the ear, eyes, and gastrointestinal tract as in Ref. [4]. Major phenotypes of developing cutaneous melanoma are pale white, red hair, which is most often found in the white population [6]. A medical expert, dermatologist helps in identifying and treating the skin diseases. Whenever they find rashes in the skin, a procedure called biopsy (removal of the tissue) has been taken from the specific site. Some of the techniques used in Ref. [6] skin biopsy is punch biopsy, shave biopsy, saucerization biopsy, wedge biopsy, incisional biopsy, excisional biopsy based on the risk of skin cancer.

Some of the risk [5] due to biopsy is bleeding, scarring, pain, and infection. It is considered to be an easy, safe, and simple procedure as in Ref. [6] before the arrival of dermatoscope. The dermatologist should have a deep understanding of the complications and skills to treat the problems due to biopsy. In order to overcome the risk due in manual biopsy, an automatic analysis of dermoscopic images are taken [16]. Dermoscope, a non-invasive technique [7] achieves 49% while diagnosing when compared without using it. Different dermoscopic techniques are polarized light dermoscopy, multispectral dermoscopy, and UV dermoscope as in Ref. [8] that provide better visualization of specific site. The light penetrates deep into the skin and so the pigmentation can be viewed clearly.

We cannot store the biopsy for a long time, but dermoscopic images are stored along with the help of software as in Ref. [9]. In the past era, to diagnose the medical images computer-aided diagnostic (CAD) system has been used. In the past decades, CAD has the capability to identify and the lesions in the skin as in Ref. [10]. Some of the challenges that exist in CAD are lack of standardized methods and so does not comparable with physician manual procedure as in Refs. [12, 13]. A technique that the system can learn automatically from the features it recognizes. The technique will combine multiple classifiers as it renders more performance as in Ref. [11] and made a huge gap to predict and quantify the patterns [15]. The accuracy degrades due to data labeling and feature engineering. In recent era, deep learning (DL) provides high effectiveness when large scale of sample is used. This technique overcomes the problem of overfitting as in Ref. [14]. It supports to take patches in 2D or 3D [15], which helps to classify the images as the target.

This chapter is organized as follows. First the dermoscopy and their types of images are described in Section 10.2. Secondly, categories of melanoma and the dataset availability are discussed in Sections 10.3 and 10.4. Third, techniques used in medical images, including machine learning and DL are described in Section 10.5. Fourth, steps proceeded to process the medical images are presented in Section 10.6. In Section 10.7, architectures such as AlexNet, U-Net, and Inception are described. Finally, in Section 10.8, parameters to calculate the performance are discussed.

10.2 DERMOSCOPE

A tool used to visualize the skin lesions, i.e., clinical patterns that are not visible to the naked eye is termed as dermoscope (a microscope) [39]. Its main principle of visibility is trans-illuminous and magnification through fluids. The significant feature of dermoscopy is to identify the skin lesions clearly. In order to get clear images three dermoscopic methods [8] are used, they are polarized contact, non-polarized contact and standard contact.

Non-polarized contact [49] has speciality to visualize superficial layer that are bluish-white and gray in color. In this method, white color cannot be visualized and predicted easily. Polarized method has the capability to visualize [49] the papillary (inner) layer of the skin. It is more compatible to view both pink and red color. More than these white crystalline color lesions are also visible.

The assurance can be made by researchers after two-step process in which first step gives difference between melanocytic and non-melanocytic lesions. Second step helps us to categorize the types of melanoma. Most of the dermoscope have the magnification [8] of around ×10 to ×20. This apparatus has been used primarily in medical fields to predict the melanoma disease. This tool simplifies the procedure and also has it in affordable price available in the market.

10.3   CATEGORIES OF MELANOMA

Melanoma (a type of skin cancer) grows deeply inside the hypodermis of the skin as it can be seen at the topmost layers (epidermis) of the skin. It will destroy the nearby tissues, thereby makes a challenge to the human survival. Those melanoma cells [40] can be classified into superficial spreading melanoma, acral-lentiginous melanoma, lentigo meligma melanoma, amelanotic, and desmoplastic melanoma and nodular melanoma as illustrated in Figure 10.1.

The superficial spreading melanoma is a common type of melanoma that affects the people of age 30 to 50. It simply spreads in the outermost skin without having any definite shape, size, and texture [42]. It can be diagnosed whenever we see the brown, pink, and blank patches on the upper skin.

Nodular melanoma [44] can grow deeper inside the dermis. It ruptures the skin cause bump/lump on the upper part of the skin. The patches are of black in color, as it develops in the face. Only 15% to 22% are affected by this type of cancer.

In Asia, 40–50% is diagnosed with acral lentiginous melanoma [43]. Not only the Asian bit also the African and Hispanic ancestors are also affected by this type of melanoma. Even though, it is difficult to diagnose at the initial stage, it develops easily in the feet, palms, and nails.

Desmoplastic melanoma which is also called neurotropic melanoma. It contains tumor silhouette may appear pink with collagenous matrix [45]. It can be seen under a light microscope. It uncommon which occurs less than 4% of primary cutaneous melanoma.

10.4   DATASETS

The research work can be done with the already available dataset. The dataset which is the collection of data that are publicly available for several application such as medical, hydrological, financial, and energy-related applications.

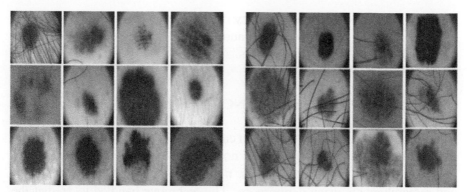

FIGURE 10.1   Dermoscopic image samples of PH² dataset.

In this survey, the dermoscopic images Figure 10.1 shows the image of the dataset that are taken into consideration to do the research. In order to solve the challenges, several medical foundations have released datasets that are publicly available. Some of the datasets available are:

1. Atlas of dermoscopy is a dataset created by Argenziano [46] consists of 768×512 pixels so-called "wet" dermoscopy which covers 1044 images of 167 non-melanocytic lesions. It has some limitations in accessing these datasets for research works.

2. The ISIC dataset (https://isic-archive.com/) composed of 13,786 dermatoscopic images as of February 2018. It is considered to be a standard source, structured, licensed, and available for image-oriented research works. It contains 12,893 images of nevi as the part of 13,786 melanoma as in Ref. [46].

3. PH² database (https://www.fc.up.pt/addi/ph2%20database.html) contains 200 dermoscopic images as in Ref. [41], among these 80 nevi, 80 atypical nevi and 40 melanoma are called by University of Porto Tecnico and Dermatology service of hospital Pedro Portugal.

4. The Danish Melanoma Database (https://www.danishhealthdata.com/find-health-data/Dansk-Melanom-Database?disallowCookies=1), a national clinical quality center which contains several patents of melanoma starting from birthmark to pre-matured stage. Those datasets can be accessible worldwide [47].

5. HAM (human against machine) 10,000 is one of the dataset that contains 10015 dermatoscopic images. Those dataset (https://www.kaggle.com/kmader/skin-cancer-mnist-ham10000) consists of 10,015 images [42] which may have both training and testing data of

different population of variety of modalities. Those datasets can be used in all machine learning and DL, and their comparison has been made with human researchers.

10.5 TECHNIQUES IN SKIN CANCER DETECTION

Before the advent of dermoscopy, skin cancer can be identified using dynamic thermal imaging (DTI), which is a non-invasive technique [48] and also provides accurate results. Nowadays, mostly in medical fields, technicians and researchers are using dermoscopy. The technique handled to predict the skin cancer [48] using images taken from dermoscope are MATLAB, machine learning and DL. Mainly particle swarm optimization (PSO), speed up robust feature (SURF) algorithms is used in MATLAB processing to segment the images. In machine learning, support vector machine (SVM), backpropagation algorithms are being used to segment and classify the skin lesions. Machine will generalize the data from its experience. Here the classifier is trained to learn [49] from the features. The techniques like [50] convolutional neural network (CNN), generative adversarial model (GAN), deep autoencoders (DAN), and convolutional autoencoders (CAE) are used in DL to diagnose the disease. While comparing with machine learning, the DL model takes less time in working with high performance GPU to analyze and predict.

10.5.1 MACHINE LEARNING IN MEDICAL IMAGING

The state of art performance has been achieved whenever the machine learning technique is applied. A basic idea, the machine learning allows machines to learn from the patterns (data) it recognizes. It helps to render medical images by computing certain features, which is believed to make good prediction [17]. To overcome the problem in traditional technique, SVM like machine learning algorithm train the machine to diagnose correctly. Machine learning has now been applied in order to classify the patterns, feature by the way it reduces the execution time.

The old imaging techniques are not successful in managing the problems whenever these techniques identify that the data is missing. In order to overcome the problem in imaging techniques, machine learning algorithms are used, which shows enhancing performance [18] by making using of train and test dataset. Most often, the performance depends on the accuracy of the machine that predicts and extracts the features. The machine-learning algorithm helps to reduce the manual processing of human workload [19] as it provides assurance. Both DL and machine learning researchers are racing in exploring the techniques with state of the art.

10.5.2 DEEP LEARNING (DL) IN MEDICAL IMAGING

Our staple goal is to familiarize the new techniques in DL that compensate and also to overcome the problem in ML. The issue in machine learning is limited amount of dataset. This can be solved by enlarging technique called augmentation as in Ref. [20]; i.e., label preserving and transformation. Computational power is also one of the challenges in ML, which can be easily handled by the fast-improving DL methods as in Ref. [1]. In terms of DL, four types of learning strategies handled [21] are:

- Supervised learning;
- Unsupervised;
- Semi-supervised/partially supervised; and
- Reinforcement is handled.

In case DL deals with the complex neural network of multiple layers as input, hidden, and output layers which have been designed to make decision. It is expected that by 2021 [22], much more investment will be spent in medical analysis in par with DL. It implies that the computational unit takes multiple inputs and then process in a non-linear fashion to produce accurate output.

10.6 STEPS TO IDENTIFY SKIN CANCER

Splendid enhancements of DL over machine learning draw an attention to explore the possibilities of medical images acquired using dermoscopy. Hence, we elaborate on how well the DL works well within medical images through the following: data augmentation, image pre-processing, and segmentation.

10.6.1 DATA AUGMENTATION

Some problem may exist in data while dealing with machine learning techniques. Both the training and testing data would have insufficient amount

[23] of data or the class of unbalanced/ unsupervised data need to analyze. Those types of data can be deal with in a specific way called data augmentation, which enhances the size and improves the quality of data to be used in training and testing dataset. There are two methods involved with data augmentation [24, 25], and they are feature space augmentation and generative adversarial network (GAN) augmentation.

10.6.1.1   FEATURE SPACE AUGMENTATION

It is one of the data augmentations based on DL technique. Noise, interpolation, and extrapolation are the features to be tested in augmentation which are discussed by DeVries and Taylor. Testing can be done by extrapolating nearest neighbors in a sample to generate the new set of samples in the way the challenges are overridden by reducing the lower-dimensional representation [26] in topmost layers of CNN. Auto-encoders are implemented in performing the feature augmentation technique.

10.6.1.2   GENERATIVE ADVERSARIAL NETWORK (GAN) AUGMENTATION

A powerful tool to manipulate the images as like text to image transformation, generating high resolution, image-to-image synthesis can be done using GAN augmentation. This would enhance the quality of data samples with the help of different auto-encoders. The image of size [height] × [width] × [color] can be sampled to a vector size of $n \times 1$. The autoencoders [23] will learn the low dimensional vector from the data points and align them within the space. Here two adversarial networks are used to generate realistic image and to compare the distinguishing features of fake and realistic image, which also minimizes the cost. This model gives satisfactory results while evaluating the generative adversarial model.

10.6.2   PRE-PROCESSING

Image acquisition is the foremost process in every medical image processing. There shouldn't be any expectation that dermoscopic images is of high quality. Artifacts are a common problem in dermoscopic images [66]; they are hair, air bubbles, and black frames which appear in human skin. Other problem is the illumination of the dermoscopic images, i.e., shading effect. Therefore, pre-processing can be done in two stages: (a) first artifact removal [67]; (b) resizing and patch extraction [68].

Before proceeding with segmentation, hair removal is an important step to get a closer visual of melanoma lesions. To detect and remove the hair from the images, a specific method used in image process called dull razor [27]. As the dull razor predicts the locations of hair using morphological operations [68] and also apply binary masking. It verifies the shape and texture using bilinear interpolation and can be smoothened by the process of adaptive median filer.

Normalization is resizing, attenuation of shading over the images. Most of the images are of uneven illumination, which makes the confusion in locating the corners. It can be solved by shading effects using the Otsu algorithm [69]. Color is one of the visual characteristics that include hue, saturation, and intensity, which reflect the emotional [28] of the image. It can be encoded in pixels (matrix) of images [3] in terms of height, width, and color. Reframing an image by cutting a pixel coordinates in the image. The cropping will reduce the size of the input image from 256 to 224. It can be done on the left and right corner of the coordinates (x, y). It is not possible to preserve the labels while cropping, but spatial dimension can be preserved while translating. The standard image [25] has coordinates (x+origin.x, y+origin.y) which would be equal to the cropped image of the coordinates (x,y). The image may also contain uncleared parts which is called occlusion. This random erasing may help the model to erase the challenges in visual features of the images. As this is one type of augmentation may overlap on other augmentation as like as stack. More accuracy can be obtained due to random erasing [33], which has been implemented.

The dermoscopic images are of various resolutions. Before being fed into a neural network, it should be resized. So, most of the images are downsampled [68] to 256×256 resolutions. Patch extraction can be performed by extracting small patches with the corresponding anatomical position. The process can be done by using non-overlapping and overlapping methods to overcome the problem of imbalancing. After the pre-processing sample, pre-processed dermoscopy images are shown in Figure 10.2 in which the size of a small patch can be chosen to be compatible [63] with specific architecture. This step gives special attention and produce good result in future steps.

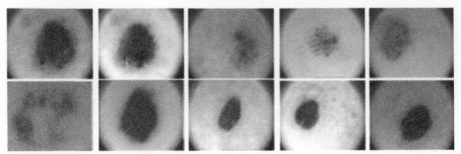

FIGURE 10.2   Pre-processed image sample of PH^2 dataset.

10.6.3   CLASSIFICATION

Pre-processed image is taken for further processing huge volume of training data is considered. Significant problem is to detect a certain area for such as purpose Deep CNN has used is used to classify. A network classifier somehow accurately classifies the images. For example, it may consist of 10,000 images, which are fed into a neural network, is being split up into training set of about 70% and 30% as its remaining for validation. Initially the input image of size 224×224 is taken to train the convolutional layer assume it may be 22 layers. The 1×1 convolution along with ReLU would considerably reduce the dimension.

As the images are of different resolution that is regulated by split, transform, and merge function. Different filter of 3×3, 5×5 with varying sizes of strides are used. The weights and bias can be calculated by each filter in the network.

Pooling layer makes use of max-pooling operation, which reduces the size of the network, and then appropriate features are analyzed from the images. Those extracted features are combined together to get loss and output classifier. Whenever any problem occurs, it automatically freezes [71] the network by initializing layer to zero. As the network [72] goes deeper it would give better accuracy. Finally, the last layer extracts the features which diagnose the categories in the dermoscopic images.

10.6.4   SEGMENTATION

Segmentation is one of the pre-requisites for quantitative and qualitative assessment [29] of melanoma lesions as it varies in [49] color, texture, and shape. Recent advancement in DL CNN dominates the medical field by segmenting dermoscopic images.

Segmentation can be divided into two types as layer and block-based [30] segmentation. Object detection can be done by layer-based segmentation. In this type, shape, depth order, and instance segmentation can be evaluated. Feature selection can be made by block-based segmentation, which would be used to create graphs based on evaluation. This type will refine the features w.r.t pixels. There are enormous methods available to segment dermoscopic images are median cut, k-means, fuzzy c-means and mean shift.

Here four metrics [70] can be used to evaluate the comparison between manual and automatic segmentation via sensitivity as the following Eqn. (1), specificity in Eqn. (2), accuracy in Eqn. (3), and similarity in Eqn. (4).

$$\text{Sensitivity} = (\text{TP/TP} + \text{FN}) \times 100\% \tag{1}$$

$$\text{Specificity} = (\text{TN/TN} + \text{FP}) \times 100\% \tag{2}$$

$$\text{Accuracy} = (\text{TP+TN/TP} + \text{FP} + \text{FN} + \text{TN}) \times 100\% \tag{3}$$

$$\text{Similarity} = (2\text{TP/2}(\text{TP} + \text{FN})) \times 100\% \tag{4}$$

FP implies pixel which represent the classified images in manual detection. FN implies pixel which represent the classified images in automatic detection. Some properties [30] are there for block segmentation, they are: (a) discontinuity and (b) similarity property. Based on the properties, the segmentation can be categorized into:

- Region-based segmentation-discontinuity property;
- Edge or boundary-based segmentation-similarity property;
- Hybrid segmentation.

10.6.4.1   REGION-BASED SEGMENTATION

Region based segmentation [31] can be done by applying clustering k-means, fuzzy clustering, and split and merge.

10.6.4.1.1   Clustering-K Means

Splitting up of images into k groups having non-overlapping groups. It can be preceded based on iterative algorithm. Minimum mean value can be considered while making boundary [33] between data points and the centroid of the cluster. It is more homogeneous rather than heterogeneous. To filter the images by clustering method, high accuracy and efficiency are needed, and hence better results can be achieved. The correlation [32] between two data points can be made by Euclidean distance using the following Eqn. (5):

$$d_{Euclidean}(x,y) = \sqrt{\sum_{i=1}^{n}(x_i - y_i)^2}$$ (5)

where; d (x, y) is the distance between data x to the center of clustery; and xi is data in n number of data.

This k-means clustering algorithm can be more effective in finding the starting point of the centroid. For each execution, different center can be taken for the same dataset and k value. Though some demerits occur, there would be improvements in obtaining local optimum in terms of time and accuracy.

10.6.4.1.2   Fuzzy Clustering

In order to predicting the objects through the method of classification referred to as clustering. Some of the different ways the clustering can be done are: (a) hard clustering and (b) fuzzy clustering.

One type of effective algorithm for image segmentation is fuzzy clustering. It provides the optimal classification to segment the data [33], which is the part of one or clustered data. It also provides the solution to multi objective optimization, which would minimize the distance to reach the objective. Of most of the algorithms, Fuzzy C-means provides good solution in segmenting the noisy image. The research work has been made on different images [34] as such: (a) Gauss noise, (b) salt noise, and (c) pepper noise. This fuzzy clustering acquires better computational speed and good segmentation when compared with other algorithms.

The cluster center [33] can be obtained by minimizing the minimized by using the following Eqn. (6):

$$J_{FCM} = \sum_{i=1}^{N}\sum_{j=1}^{c}u_{i.j}^{m}d^2(x_i,v_j), 1 < m < \infty$$ (6)

Here m implies blur exponent; $X = \{x_1, x_2, \ldots x_N\}$ that contains N pixels; V $= (v_1, v_2, \ldots v_C\}$ is the center of the cluster; $d_2(x_i, v_j)$ is the Euclidean distance of the object x_i; and u_{ij} is the degree of membership.

Hence the clustering can be exploited in medical and security-oriented research work.

10.6.4.1.3 Split and Merge

For all the images, the splitting the pixels and merging the two adjacent regions can be done by segmentation process. It is a top-down approach and proceeds reverse of region growing [31, 36] technique. It splits up the pixels of the whole image in homogenous factor into the non-homogenous segregated image. Bimodality approach can be applied in splitting the image, which is having shape and size as limiting factor. User can divide the kernel points into several sub-regions [35] into a set of unconnected regions, and then integration can be made based on quadtree structure.

10.6.4.2   SEMANTIC SEGMENTATION

Rather than putting a bounding box to the image, it labels each pixel in a medical image. Hence the semantic segmentation is called denser prediction. In medical image, lesions will be segmented as per corresponding classes. Here class symbolizes [37] the color in the patches as pink, black, and brown. It shows the high resolution.

It gives the solution to coarse and fine inference. In medical images, lesions prediction is transformed from region to pixel-based prediction. Each pixel in medical image will be classified as class labels. The output image would be high-resolution [38] make the dense prediction to each pixel in the image.

10.7   ARCHITECTURES

In 1988, Fukushima proposed the CNN structure. However, previous techniques are not compatible [52] for heavy computational load. Now-a-days, most of the researchers inspired to use the deep CNN. The CNN consists of several features that may enhance the performances. Moreover, several enhancements have been made in CNN learning architectures which consist of features [51] as large, heterogeneous, and multiple classes of solution to the problem. Enormous data usage and improvement in GPU utilization motivated the researcher in CNN, and thus interesting deep CNN models has been exploited. Each of the CNN models have unique facet [51] some of those are design pattern, connectivity, processing units and optimization strategies. Based on enhanced features, CNN can be categorized as AlexNet, GoogLeNet, ResNet, and U-Net are described.

10.7.1 ALEXNET

In 2012, Alex Krizhevsky, Geffrey, and Ilya Sutskever made the AlexNet architecture [53], which is a basic CNN. Among the DL architecture, AlexNet is the winner of the image net competition [54] conducted in 2012. Moreover, it consists of eight layers, which is deeper and wider than LeNet (5 layers) that provides good accuracy and capability to generate images with different resolutions [51]. It consists of cascaded layers of five convolutional layers, pooling layer, and three fully-connected layer (FC layer). In these two, one fully connected hidden layer and one fully connected output layer are there. AlexNet architectural design is depicted in Figure 10.3.

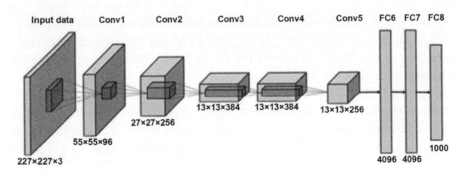

FIGURE 10.3   The AlexNet architecture.

To address the challenges [54] in hardware, two NVIDIA GTX 580 GPU are used as they enlarge the network size and reduce the time to train than one GPU. Recognition, classification, and segmentation can be made using DL, which shows the quantum leap than machine learning approaches. The fully connected output layer [54] is connected to SoftMax, which provides the classification over many class labels.

In AlexNet [52] input is fed into convolutional layer which performs the convolution operation and max-pooling operation along with receptive filters are used. Those filters are of different sizes (11×11 and 5×5) used at the initial stage [51]. Most of the max pooling operations uses filter of size 3×3 and stride of size 2. AlexNet is being triumphant due to ReLU non-linearity and dropout regularization [53]. ReLU achieves state of art results [55] due to faster learning and better performance while comparing with sigmoid and tanh activation functions. Then Eqn. (1) used by ReLU activation function carry out the threshold operations [56]. This operation shown in Eqn. (7) is done in hidden layers and also in output layers of DL network:

$$f(x) = \max (0, x) \tag{7}$$

As this ReLU is a half-wave rectifier function avert the overfitting problem and so enhances the convergence rate [51]. Even though both AlexNet and LeNet are almost similar in structure, there are some differences exist in the model.

The second, third, fourth, and fifth layer performs similar operations but different filter size. In this network, third, fourth and fifth layer uses 3×3 filters but only second layer uses 2×2 filters. Finally, the fully connected output layer is SoftMax and performs dropout operation at the end.

Whenever the input of size 224×224×3 is fed into [52] convolutional layer, it produces an output of size 55×55×96. In a similar way, the next layer of operation can be preceded. Thus, the way AlexNet gives the remarkable importance to the natural dataset and remote sensing dataset and provides a new way to researchers in the CNN model.

10.7.2   GOOGLENET

Christian Szegedy developed a model called GoogLeNet. This model wins Image Net Large Scale Visual Recognition [57] Challenge (ILSVRC-2015). The name GoogLeNet came from LeNet and is the epitome [58] of Inception architecture. This network make use of small blocks of network with the main objective is to achieve the better performance [51] by reducing the computational complexity [52] by deepening the convolutional layer. Despite of the fact, GoogLeNet comprises of 22 layers than any other network. The state of art accuracy has been achieved due to the inception layer which is aligned in stacked form [52]. The general architecture of GoogLeNet is shown in Figure 10.4.

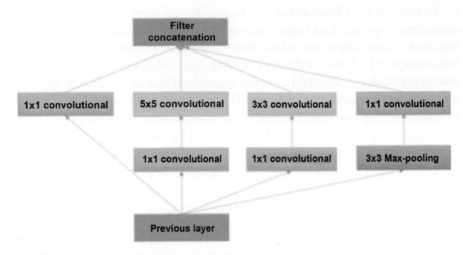

FIGURE 10.4   GoogLeNet with inception layer.

The proposed model uses convolutional kernels of size 1×1, which would reduce the overall dimension and rectified linear activation. As per the model, the independent layers of kernel size 1×1, 5×5, 3×3 and max-pooling takes the input. In each step, the input data is fed into a deeper layer to extract the appropriate features by reducing the channel size.

The feature extraction can be made by encompassing the split, transform, and merge techniques and also to address the problem of same or different resolution of images. The max-pooling layer achieves the top-1 [58] accuracy of 0.6%. Moreover, parameters are reduced [51] from 138 million to 4 million. While comparing is much more lesser than AlexNet. A maximum of 1024 is there in FC layer with rectified linear activation function. The fully connected output layer along with SoftMax classifies the images of the same 1000 classes. Furthermore, the convergence rate can be accelerated by auxiliary learners. Some limitation exist in GoogLeNet are sometimes even few information can be lost and the topology can be customized for each module.

10.7.3   RESNET (RESIDUAL NETWORK)

Kaiming He [52] proposed the ResNet model and was owned by Microsoft. In 2016, ResNet won both the challenges of ILSVRC and COCO challenge [59]. The convolutional layer and sub-sampling layer together form the

ResNet. It is also called a traditional feed-forward network [52]. ResNet is 20 times intense than AlexNet and eight times than that of VGG [54]. ResNet, which implies layers are stacked together as it vanishes the gradient problems. The building block of ResNet architecture is depicted in Figure 10.5. It consists of layers in terms [52] of 34, 50, 101, 152, and even more than 1200. The gradient problem would get diminished whenever the network is deeper by making use of the backpropagation technique.

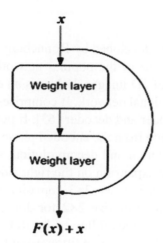

FIGURE 10.5   Building block of ResNet architecture.

However, ResNet can be modified by only using the FC layer and works based on residual network (ResNet). Based on the architecture, ResNet operations can be modified. The output of ResNet [52] depends upon the previous layers. Whenever the two stacks of layers are build would build the following Eqn. (8):

$$H(x) = F(x, \{Wi\}) + x \qquad (8)$$

Here x implies input and H(x) is the output and F (x, {Wi}) is the residuals [60].

The identity mapping can be made by formulating a shortcut connection as F(x) +x with the help of feed-forward neural network (FNN). Before forwarding the output to the next layer, many convolutions operations of variant kernel size, batch normalization (BN) and activation function ReLU can be carried out.

Whenever the training gets initiated, the network are set to zero weights and perform identity mapping. Whenever it identifies the error, the weights in the network get to be changed. It is easy to deploy and show the good results as it can be compared to GoogLeNet. However, some limitations exist in ResNet are cost to copy and modeling the network. It achieves top-5 error [58] of 3.57% and provides good accuracy than GoogLeNet.

10.7.4   U-NET

The U-Net architecture is developed by Ronneberger Fischer, and Brox [62] Olaf Ronneberger. It is one of the popular networks [61] used to perform semantic segmentation of 2D images. U-Net is an extended [63] version of fully connected convolutional network. It comprises of both contraction and expansion phase as encoder and decoder [63]. It performs spatial extraction and construction of feature from the encoded phase. In another way, it is the amalgamation of convolution and deconvolution layer. Contraction can be preceded in left side and expansion on the right side. The contraction phase extracts the features by using two 3×3-convolution operations [64] along with max-pooling operation of size 2×2 for downsampling. The sequence process is repeated 3 to 4 times followed by Ref. [62] rectified linear unit (ReLU). The ReLU function can be written using Eqn. (9):

$$f(x) = \max(0, x) \tag{9}$$

On the expansion (decoder) side transposed convolution operation of size 2×2 can be proceeded. The information from the contraction phase is combined together in the expansion phase. Skip convolution (ReLU) operation to retrieve the lost information. The U-Net architecture design is shown in Figure 10.6.

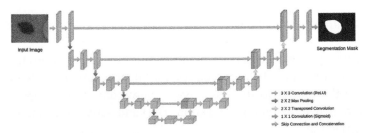

FIGURE 10.6   The U-net architecture.

Whenever the input [64] is fed into U-Net, it first produces the 64-channel along with Conv+BN+ReLU operation. It subsequentially generates 128, 256, 512, and 1024 along with max pool (Conv+BN+ReLU). Next, four consecutive up convolution can be proceeded along with (Conv+BN+ReLU) up to 512, 256 from 1024. The intermediate features [65] are generated by using skip connection on the decoder side.

10.7.5 INCEPTION V2

The inception v2 upgrade the accuracy through the BN process. Hence in the v2 module, the computation overhead has reduced gradually. The convolution operation varies by replacing 5×5 in terms of two times of 3×3 convolutions, thereby reduces the parameters, and also the network can move one step ahead than inception v1. The 3×3 convolution can be stacked as like as a layer to improve the performance [77], and 5×5 convolutions would be 2.78 times lower in computational speed than 3×3 convolutions. As this module goes deeper, in order to move fast, filter banks have been introduced as it would reduce the bottleneck. The efficiency can be computed would be (2 × (3×3))/ (5×5) or 27.3%. The inception of v2 is depicted in Figure 10.7.

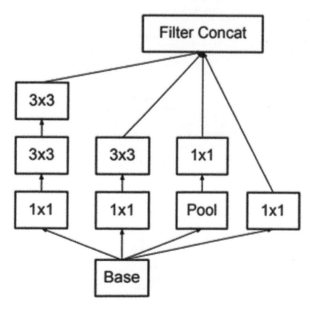

FIGURE 10.7 Architecture diagram for inception v2.

10.7.6 INCEPTION V4-RESNET

It can also be called a deduction block is shown in Figure 10.8. In ResNet, some problems may occur, which is due to degradation that may split into two phases as saturation accuracy and degradation accuracy. The degradation problem can be solved by passing the input images [78] through the stem. Here stem implies a number of convolution operations performed. The operations of inception-ResNet are similar to inception v3. Similar to other modules' inception versions, ResNet takes input, which is the output to successive convolution operations.

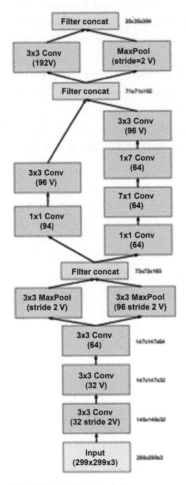

FIGURE 10.8   Inception v4 ResNet architectural diagram.

10.7.6.1   CHALLENGES OF DL

There are several challenges [52] exist in DL. The DL provides good accuracy to calculate the location and diagnose the disease correctly. However, in 2012 AlexNet solves the data augmentation problem which provides internal representation to improve performance.

It supports the efficient smart devices somehow is employs the powerful GPU for the effective CNN automation. Example for DL in such applications is smart cities, law enforcement [73–75]:

- Scalability of CNN approach;
- Multi-task and transfer learning (generalization) or multi-module learning. This means learning from different domains or with different models together.

Those challenges are seriously among the DL community, and solution are also provided by them.

10.7.6.2   WHY DEEP LEARNING?

1. It is one of the universal approaches [52] that can be exploited in any application domain.
2. Deep learning is robust as it automatically extracts the features whenever any variations are present in data.
3. Generalization.
4. Scalability.

The DL approach is highly scalable. Whenever the network contains 1202 layers and is supports supercomputing scale.

10.8   PERFORMANCE INDICES

The performance can be made using trained models in DL framework along with the help of PyTorch framework. Major challenge exists in object with low resolution. This architecture is cheaper to overcome the challenges [76] by making slight adjustments to the existing input images. All the images may have a height and width in RGB color coordinates. For example:

- NASNet contains 331 pixels;
- Inception ResNet-v2, Inception-v3, Inception-v4 contains 229 pixels.

The accuracy, computational, and model complexity, inference time, and memory usage can be evaluated to compute the performance of the DL framework:

1. **Accuracy:** Average prediction can be implemented on a dataset of melanoma. For which the Top-1 and Top-5 accuracy can be evaluated by computing the image classification [76] on different datasets. The performance can be evaluated for AlexNet, Inception v1, v2, and ResNet architecture in terms of Top-1 and Top-5 accuracy.
2. **Complexity:** Both the model and computational complexity can be estimated using a deep neural network model. The size of the model is specified in terms of MB. The computational cost is calculated using the floating-point operation (FLOPs). Separate Flop can be added and make the convolutions to be bias-free. This type of complexity [76] gives the understanding in estimating the GPU memory and number of FLOPS in convolutions.
3. **Memory:** Measuring the total memory consumption, which contains information about both memory allocated and deallocated space. For which batch processing can be done for batch sizes: 1, 2, 4, 8, 16, 31, and 64. The network model can be formed by using both memory models [76].
4. **Time:** The time can be calculated in terms of milliseconds along with batch size. For all the images, i.e., both training and testing data set can be considered in reporting statistical [76] evaluation for all DNN models.

KEYWORDS

- **batch normalization**
- **convolutional autoencoders**
- **deep autoencoders**
- **dynamic thermal imaging**
- **floating-point operation**
- **generative adversarial network**

REFERENCES

1. Ferlay, J., Colombet, M., Soerjomataram, I., Mathers, C., Parkin, D. M., Pineros, M., Znaor, A., & Bray, F., (2019). Estimating the global cancer incidence and mortality in 2018, GLOBOCAN sources and methods. *Int. J. Cancer, 144*, 1941–1953.
2. Trotter, S. C., Sroa, N., Winkelmann, R. R., Olencki, T., & Bechtel, M., (2013). A global review of melanoma follow-up guidelines. *J. Clin. Aesthet. Dermatol., 6*, 18–26.
3. Wadhawan, T., Situ, N., Rui, H., Lancaster, K., Yuan, X., & Zouridakis, G., (2011). Implementation of the 7-point checklist for melanoma detection on smart handheld devices. In: *Proc. 2011 IEEE Eng. Med. Biol. Soc.* (pp. 3180–3183).
4. McCourt, C., Dolan, O., & Gormley, G., (2014). Malignant melanoma: A pictorial review. *Ulster Med. J., 83*(2), 103–110.
5. Riker, A. I., Zea, N., & Trinh, T., (2010). The epidemiology, prevention, and detection of melanoma. *Ochsner J., 10*(2), 56–65.
6. Nischal, U., Nischal, K. C., & Khopkar, U., (2008). Techniques of skin biopsy and practical considerations. *J. Cutan. Aesthet. Surg., 1*, 107–111.
7. Kittler, H., Pehamberger, H., Wolff, K., & Binder, M., (2002). Diagnostic accuracy of dermoscopy. *Lancet Oncol., 3*, 159–165.
8. Kaliyadan, F., (2016). The scope of the dermoscope. *Indian Dermatol Online J., 7*, 359–363.
9. Sonthalia, S., & Kaliyadan, F., (2019). *Dermoscopy Overview and Extra Diagnostic Applications*. Stat Pearls.
10. Albahar, M. A., (2019). Skin lesion classification using convolutional neural network with novel regularizer. *IEEE Access, 7*, 38306–38313.
11. Hekler, A., Utikal, J. S., Enk, A. H., et al., (2019). Superior skin cancer classification by the combination of human and artificial intelligence. *Eur. J. Cancer, 120*, 114–121. doi: 10.1016/j.ejca.2019.07.019.
12. Doi, K., (2007). Computer-aided diagnosis in medical imaging: historical review, current status and future potential. *Comput. Med. Imaging Graph, 31*, 198–211.
13. Jorritsma, W., Cnossen, F., & Van, O. P. M. A., (2015). Improving the radiologist-CAD interaction: Designing for appropriate trust. *Clinical Radiology, 70*(2), 115–122.
14. Russakovsky, O., Deng, J., Su, H., Krause, J., Satheesh, S., Ma, S., Huang, Z., Karpathy, A., Khosla, A., Bernstein, M., et al., (2014). *ImageNet Large Scale Visual Recognition Challenge*. arXiv:1409.0575.
15. Shen, D., Wu, G., & Suk, H., (2017). Deep learning in medical image analysis. **Annu. Rev. Biomed. Eng.,** *19*, 221–248.
16. Fechete, O., Ungureanu, L., Şenilă, S., Vornicescu, D., Dănescu, S., Vasilovici, A., Candrea, E., et al., (2019). Risk factors for melanoma and skin health behavior: An analysis on Romanian melanoma patients. *Oncol Lett., 17*, 4139–4144.
17. Lundervold, A. S., & Lundervold, A., (2019). An overview of deep learning in medical imaging focusing on MRI. *Z. Med. Phys., 29*, 102–127.
18. LeCun, Y., Bengio, Y., & Hinton, G., (2015). Deep learning. *Nature, 521*(28), 436–444.
19. Narayanan, B. N., Djaneye-Boundjou, O., & Kebede, T. M., (2016). Performance analysis of machine learning and pattern recognition algorithms for Malware classification. In: *Proceedings of the 2016 IEEE National Aerospace and Electronics Conference and Ohio Innovation Summit, (NAECON-OIS '16)* (pp. 338–342). USA.

20. O' Mahony, N., et al., (2019). Deep learning vs. *Traditional Computer Vision Advances in Computer Vision* (pp. 128–144). Springer Nature, Switzerland AG.

21. Alom, M., Tha, T., Yakopcic, C., Westberg, S., Sidike, P., Nasrin, M., Hasan, M., Essen, B., Awwal, A., & Asari, V., (2019). A state-of-the-art survey on deep learning theory and architectures. *Electronics, 8*, 292.

22. Razzak, M. I., Naz, S., & Zaib, A., (2017). *Deep Learning for Medical Image Processing: Overview, Challenges and Future.* arXiv preprint arXiv:1704.06825.

23. Mikoajczyk, A., & Grochowski, M., (2018). Data augmentation for improving deep learning in image classification problem. In: *2018 International Interdisciplinary PhD Workshop (IIPhDW)* (pp. 117–122).

24. Bisla, D., Choromanska, A., Stein, J. A., Polsky, D., & Berman, R., (2019). *Towards Automated Melanoma Detection with Deep Learning: Data Purification and Augmentation.* arXiv:1902.06061 2019 [online] Available at: https://arxiv.org/abs/1902.06061 (accessed on 18 December 2020).

25. Shorten, C., & Khoshgoftaar, T., (2019). A survey on image data augmentation for deep learning. *J. Big Data, 6*, 60.

26. DeVries, T., & Taylor, G. W., (2017). Dataset augmentation in feature space. In: *Proceedings of the International Conference on Machine Learning (ICML), Workshop Track.*

27. Somnathe, P. A., & Gumaste, P. P., (2015). A review of existing hair removal methods in dermoscopic images. *IOSR Journal of Electronics and Communication Engineering (IOSR-JECE), 1*, 73–76.

28. Barata, C., et al., (2015). Improving dermoscopy image classification using color constancy. *IEEE J. Biomed. Health Inform., 19*, 1146–1152.

29. Olugbara, O. O., Taiwo, T. B., & Heukelman, D., (**2018**). Segmentation of melanoma skin lesion using perceptual color difference saliency with morphological analysis. *Math. Probl. Eng., 19.*

30. Kaur, D., & Kaur, Y., (2014). Various image segmentation techniques: A review. *International Journal of Computer Science and Mobile Computing (IJCSMC), 3*, 809–814.

31. Kaur, M., & Goyal, P., (2015). A review on region-based segmentation. *International Journal of Science and Research (IJSR), 4*, 3194–3197.

32. Kahkashan, K. S., (2013). A comparative study of K means algorithm by different distance measures. *International Journal of Innovative Research in Computer and Communication Engineering*, 2443–2447.

33. Yang, Y., & Huang, S., (2007). Image segmentation by fuzzy c-means clustering algorithm with a novel penalty term. *Comput. Inf., 26*(1), 17–31.

34. Bhosale, N. P., & Manza, R. R., (2014). Analysis of effect of Gaussian, salt and pepper noise removal from noisy remote sensing images. In: *Second International Conference on Emerging Research in Computing, Information, Communication and Applications (ERCICA).* Elsevier. ISBN: 9789351072607.

35. Anju, B., & Aman, K. S., (2017). Split and merge: A region-based image segmentation. *International Journal of Emerging Research in Management and Technology, 6*(8), ISSN: 2278-9359.

36. Chaudhuri, D., & Agrawal, A., (2010). Split-and-merge procedure for image segmentation using bimodality detection approach. *Defense Sci. J., 60*(3), 290–301.

37. Papandreou, G., Chen, L. C., Murphy, K., et al., (2015). *Weakly-and Semi-Supervised Learning of a DCNN for Semantic Image Segmentation [J].* arXiv preprintarXiv:1502.02734.
38. Guo, Y., Liu, Y., Georgiou, T., & Lew, M. S., (2017). A review of semantic segmentation using deep neural networks. *International Journal of Multimedia Information Retrieval.*
39. Nischal, K. C., & Khopkar, U., (2005). Dermoscope. *Indian J Dermatol Venereol Leprol., 71,* 300–303.
40. Pamberger, H., Steiner, A., & Wolff, K., (1987). *In vivo* epiluminescence microscopy of pigmented skin lesions-I. Pattern analysis of pigmented skin lesions. *J. Am. Acad. Dermatol., 17,* 571–583.
41. Mendonca, T., Ferreira, P. M., Marques, J. S., Marcal, A. R., & Rozeira, J., (2013). PH2: A dermoscopic image database for research and benchmarking. *Conf Proc IEEE Eng. Med. Biol. Soc.,* 5437–5440.
42. Kienstra, M. A., & Padhya, T. A., (2005). Head and neck melanoma. *Cancer Control., 12*(4), 242-24716258496.
43. Cohen, L. M., (1995). Lentigo maligna and lentigo maligna melanoma. *J. Am. Acad. Dermatol., 33*(6), 923–936; quiz 937-40. [PubMed: 7490362].
44. Erkurt, M. A., Aydogdu, I., Kuku, I., Kaya, E., & Basaran, Y., (2009). Nodular melanoma presenting with rapid progression and widespread metastases: A case report. *J. Med. Case Rep., 3,* 50.
45. Chen, L. L., Jaimes, N., Barker, C. A., Busam, K. J., & Marghoob, A. A., (2013). Desmoplastic melanoma: A review. *J. Am. Acad. Dermatol., 68,* 825–833.
46. Argenziano, G., et al., (2000). *Interactive Atlas of Dermoscopy (Book and CDROM).* Edra Medical Publishing and New Media.
47. Dansk Melanoma Database [webpage on the Internet] National Årsrapport, (2014). Available at from: https://www.sundhed.dk/content/cms/30/57130_%C3%A5rsrapport_melanomer_2014_endelig.pdf (accessed on 18 December 2020).
48. Aljawawdeh, A., Imraiziq, E., & Aljawawdeh, A., (2017). Enhanced k-mean using evolutionary algorithms for melanoma detection and segmentation in skin images. *International Journal of Advanced Computer Science and Applications, 8*(12), 477–483.
49. Mishra, N. K., & Celebi, M. E. (2016). *An Overview of Melanoma Detection in Dermoscopy Images Using Image Processing and Machine Learning. arxiv.org: 1601.07843.*
50. Munir, K., Elahi, H., Ayub, A., et al., (2019). Cancer diagnosis using deep learning: A bibliographic review. *Cancers,* 11.
51. Khan, A., Sohail, A., Zahoora, U., & Qureshi, A. S., (2019). *A Survey of the Recent Architectures of Deep Convolutional Neural Networks.* [online] Available at: https://arxiv.org/abs/1901.06032 (accessed on 18 December 2020).
52. Alom, M. Z., Taha, T. M., Yakopcic, C., Westberg, S., Hasan, M., Esesn, B. V., & Asari, V. K., (2018). *The History Began from AlexNet: A Comprehensive Survey on Deep Learning Approaches.* arXiv. [Online]. Available at: http://arxiv.org/abs/1803.01164 (accessed on 18 December 2020).
53. Han, X. B., Zhong, Y. F., Cao, L. Q., & Zhang, L. P., (2017). Pre-trained AlexNet architecture with pyramid pooling and supervision for high spatial resolution remote sensing image scene classification. *Remote Sens., 9*(8), 22. Article ID: 848. https://doi.org/10.3390/rs9080848.

54. Krizhevsky, A., Sutskever, I., & Hinton, G. E., (2012). *ImageNet Classification with Deep Convolutional Neural Networks.* [Online]. Available at: https://doi.org/10.1145/3065386 (accessed on 18 December 2020).

55. Nair, V., & Hinton, G. E., (2010). *Rectified Linear Units Improve Restricted Boltzmann Machines,* 807–814. Haifa. [Online]. Available at https://www.cs.toronto.edu/~fritz/absps/reluICML.pdf (accessed on 18 December 2020).

56. Dahl, G. E., Sainath, T. N., & Hinton, G. E., (2013). Improving deep neural networks for LVCSR using rectified linear units and dropout. In: *International Conference on Acoustics, Speech and Signal Processing.* IEEE.

57. Nwankpa, C., et al., (2018). *Activation Functions: Comparison of Trends in Practice and Research for Deep Learning.* arXiv preprint arXiv:1811.03378.

58. Szegedy, C., Liu, W., Jia, Y., Sermanet, P., Reed, S., Anguelov, D., Erhan, D., et al., (2015). *Going Deeper with Convolutions.* In CVPR.

59. He, K., Zhang, X., Ren, S., & Sun, J., (2016). Deep residual learning for image recognition. In: *Proceedings of the 2016 IEEE Conference on Computer Vision and Pattern Recognition (CVPR).* https://doi.org/10.1109/CVPR.2016.90.

60. Li, B., & He, Y., (2018). An improved ResNet based on the adjustable shortcut connections. *IEEE Access, 6,* 18967–18974.

61. Lundervold, A. S., & Lundervold, A., (2019). An overview of deep learning in medical imaging focusing on MRI. *Zeitschrift für Medizinische Physik, 29*(2), 102–127.

62. Ronneberger, O., Fischer, P., & Brox, T., (2015). U-net: Convolutional networks for biomedical image segmentation. In: *Lecture Notes in Computer Science (Including Subseries Lecture Notes in Artificial Intelligence and Lecture Notes in Bioinformatics).*

63. Jaworek-Korjakowska, J., (2018). A deep learning approach to vascular structure segmentation in dermoscopy color images. *BioMed. Research International, 2018.*

64. Ibtehaz, N., & Rahman, M. S., (2019). *Multiresunet: Rethinking the U-Net Architecture for Multimodal Biomedical Image Segmentation.* arXiv preprint arXiv:1902.04049.

65. Zhao, X., Yuan, Y., Song, M., Ding, Y., Lin, F., Liang, D., & Zhang, D., (2019). Use of unmanned aerial vehicle imagery and deep learning UNet to extract rice lodging. *Sensors, 19,* 3859.

66. Madooei, A., Drew, M. S., Sadeghi, M., & Atkins, M. S., (2012). Automated preprocessing method for dermoscopic images and its application to pigmented skin lesion segmentation. In: *Proceedings of the 20th Color and Imaging Conference: Color Science and Engineering Systems, Technologies, and Applications* (pp. 158–163).

67. Meskini, E., Helfroush, M. S., Kazemi, K., & Sepaskhah, M., (2018). A new algorithm for skin lesion border detection in dermoscopy images. *J. Biomed. Phys. Eng., 8*(1), 117–126.

68. Zafar, K., Gilani, S. O., Waris, A., Ahmed, A., Jamil, M., Khan, M. N., & Sohail, K. A., (2020). Skin lesion segmentation from dermoscopic images using convolutional neural network. *Sensors, 20*(6), 1601.

69. Otsu, N., (1975). A threshold selection method from gray-level histograms. *Automatica, 11,* 23–27.

70. Garnavi, R., et al., (2011). Border detection in dermoscopy images using hybrid thresholding on optimized color channels. *Computerized Medical Imaging and Graphics, 35*(2), 105–115.

71. Salido, J. A. A., & Ruiz, C., (2018). Using deep learning to detect melanoma in dermoscopy images. *Int. J. Mach Learn Comput., 8*(1), 61–68.

72. Sudha, K. K., & Sujatha, P. (2019). A qualitative analysis of GoogLeNet and AlexNet for fabric defect detection. *Int. J. Recent Technol. Eng. 8,* 86–92.

73. Hinton, G., et al., (2012). Deep neural networks for acoustic modeling in speech recognition: The shared views of four research groups. *IEEE Signal Process. Mag., 29*(6), 82–97.

74. Lu, H., et al., (2017). Wound intensity correction and segmentation with convolutional neural networks. *Concurr. Comput. Pract. Exp., 29*(6), e3927.

75. Hinton, G. E., Krizhevsky, A., & Wang, S. D., (2011). Transforming auto-encoders. In: *International Conference on Artificial Neural Networks* (pp. 44–51).

76. Bianco, S., Cadene, R., Celona, L., & Napoletano, P., (2018). Benchmark analysis of representative deep neural network architectures. *IEEE Access, 6,* 64270–64277.

77. Girshick, R., (2015). Fast R-CNN. *Computer Science.*

78. Sultana, F., Sufian, A., & Dutta, P., (2018). Advancements in image classification using convolutional neural network. In *2018 Fourth International Conference on Research in Computational Intelligence and Communication Networks (ICRCICN)* (pp. 122–129).

CHAPTER 11

DEEP LEARNING AND ITS APPLICATIONS IN BIOMEDICAL IMAGE PROCESSING

V. V. SATYANARAYANA TALLAPRAGADA

Associate Professor, Department of ECE,
Sree Vidyanikethan Engineering College, Tirupati, Andhra Pradesh,
India, E-mail: satya.tvv@gmail.com

ABSTRACT

In recent days, systems are designed to have better classification based on input. The inputs vary based on the application that is intended, viz., retina image for detection of diabetic retinopathy (DR). Statistical data for classification based on various input parameters, but these are not limited. This chapter mainly focuses on the issues of biomedical image processing, having its roots in deep learning (DL). Existing techniques before the evolution of DL have made their mark, but its performance is limited. These techniques fail as the database (DB) size increases. In addition, various constraints need to be considered while calculating the recognition accuracy. Therefore, it is the need of the hour to find feasible solutions for improving the classification accuracy. Primarily these techniques are classified as supervised and unsupervised classification techniques. It is a well-known fact that supervised techniques have *apriori* data while the latter does not. Obviously, the former results in better accuracy, and the latter does not. In the real world scenario, unsupervised classification is more relevant as the input data tends to vary with time.

DL is the fastest-growing field that has already proven its worth in machine learning. The attempt to model large-scale data by the use of various layered networks led to the development of deep neural networks (DNN), providing an application to various fields, but not limited to viz., image classification, and pattern recognition. In general, DL has two properties, one is various layers that support non-linear processing, and the other is supervised or unsupervised learning based on the features existing on each of the layers. Before DL development of artificial neural networks (ANNs) has enthralled into science and technology, and from 2006, DL created its impact, and now it's so deep that still, scientists are learning it. An essential part of DL marks its path into optimization, a task that provides data that best fits transfer function and finds a better fitting curve. Such optimization has various applications in classification and recognition. As in the case of an Iris recognition system, as the DB size increases, the features representing a particular class of Iris must also increase, or else the system tends to fail. Hence, optimization is required to capture optimal features for proper recognition.

Another application of DL in classification and recognition is in genomic signal processing. The classification of a large set of genome data is a trivial task, and keeping in mind the classification of such an extensive data set is hectic. The systems so trained and developed must be capable enough to handle such extensive data. DL is one of the applications that help users to find a particular genome code resulting in the identification of a particular disease or ailment from the given more extensive data set. The system tends to become more complicated for the search of a particular genome sequence, given different sequences of particular organism viz., COVID-19.

This chapter discusses DL architectures that are used in biomedical image analysis viz., CNN, DBN, and RNN. Architecture is SAE, which has found its application in skin analysis, detection of organs in 4D patient information, segmentation of hippocampus from infant's brains, optic disc extraction from fundus images. DBN is another architecture that has provided various applications such as segmentation of the left ventricle from heart MRI images, identification, and segregation of various retinal diseases. DBN, DNN, and RNN are different DL architectures that are used in the analysis of the genome sequences viz., finding the splicing patterns in individual tissues, highlighting the pathogenicity of genetic variants, identification of splice junction at DNA level, and understanding of the non-coding genome, identification of miRNA precursor and targets. Protein structure is also predicted using DBN, CNN, and RNN. Various models were developed for identification of structural binding choices, and prediction of binding sites of RBPs, disordered protein, other structures, local backbone positions, the area covered by proteins.

Keeping aside the advantages and uses of DL, problems while applying DL algorithms in biomedical applications persist. Besides having achievable accuracy and better speed, the computational complexity of these algorithms also increases based on classification methods. Labeling of medical images is a hectic task that requires professional training. Instead, these images have a privacy lock that cannot be used by the general public and researchers. Therefore, getting such data is also a complicated process.

Further, getting such a large amount of data is also a trivial task. Still, now metrics for the classification process are under development. The developed metric must also be uniform in assessment for various types of data and techniques that are existing in the networks. With the availability of such large data sets, it is also necessary to analyze the developed model carefully and adapt the model based on the features, properties, or characteristics of the data. As technology is accessing more data, which is accessed by wearable sensors via smartphones, DL helps as a tool for interpreting such data, detection, prognosis, prevention, diagnosis, and therapy.

11.1   MEDICAL IMAGING

Medical imaging is a part of medicine wherein it deals with techniques that are useful for the creation of visualizations and representation of the internal parts of the human body viz., tissues, bones, muscles. Such visualization helps the doctors to identify the exact location of the organ or defective part of the body. For example, a cancerous cell can be easily identified only when visualization is possible, in particular, in-depth of the brain, if a cancerous cell is there.

A question arises that 2D imaging already exists, and is there a quite necessity of the 3D imaging?

The answer is very much clear that the 3D visualization has given better spatial feature representation in contrast to a 2D visualization. Further, such high-resolution imaging provides better improvement in decision making before going for surgery. Previous imaging techniques provide details that given a dilemmatic situation; hence, 3D imaging has evolved.

The data that is extracted from such imaging technologies will be enormous and need many processing mechanisms. In other words, the computational complexity of the algorithms that process such data will be high and can be reduced with the use of high-end systems [10]. Further, it must be noted that medical imaging results entirely depend on the quality of the images. The images that are acquired at the input stage must be of high quality. Noise consideration is a must for further processes. At this juncture, it is quite imperative to analyze the outcome of mixed noise. Such noise can be introduced at any stage between input and output. The process of reducing such noise may be heuristic. Again the process of reducing the noise along with the computational complexity of the processing algorithm used for processing medical images must not get added to this algorithm. Hence, such an algorithm needs to effective in processing as well as output. The images that are acquired will not be homogeneous. The systems that are used for acquiring the medical images may be the same. Hence, the images that are acquired will be heterogeneous. Even though they are acquired from the same patient, depending on the conditions at the time of acquisition, they may vary. In other words, the biological structure change. Thus, apriori knowledge of the structure as well as the system is required.

Figure 11.1 shows the process or building blocks involved in medical imaging. The initial phase starts with image acquisition, wherein a particular band or wavelength is selected. For example, Iris is acquired in the near-infrared region, whereas other images are acquired in the visible region. Therefore, based on the application, the acquisition needs to be selected. Figure 11.1 can be basically divided into three phases. First phase starts from acquisition to storage. The processes involved in the first stage is image acquisition, then enhance the image for further analysis, which will aid the doctor to particularly identify the ailment. Then validation is performed by the concerned radiologist whose is a subject expert and will validate subjectively. This is further stored in the database (DB) for future use. The second phase is the evaluation procedure that is carried out for further research. This is crucial. The second phase involves segmentation and validation in line with mapping functionality. Mapping here means the image is projected onto high dimensional feature space for visualization or extraction of the feature vectors. This process is generally used before the classification. This process is also called functional imaging, wherein the image the formation is carried out.

Then representation, along with segmentation, is performed, which is then validated by the radiologist again and stored in the DB for future use. The third phase is the diagnosis and proposal of treatment. This step is anyhow carried out by the doctors. However, the two phases before the third phase requires accurate formulations, hence, resulting in better diagnosis and treatment.

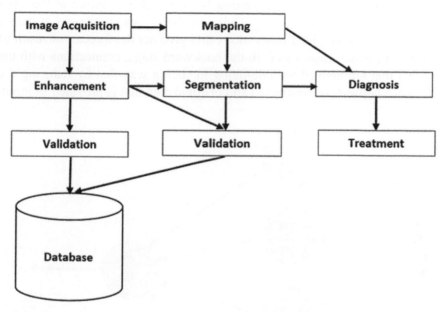

FIGURE 11.1   Medical imaging processes.

11.2   OVERVIEW OF DEEP LEARNING (DL)

DL [3–9] is derived from an artificial neural network (ANN), which is now used in the development of algorithms for machine learning, computer vision, etc. The DL models normally adopt ordered structures viz., hierarchical, that connect the layers. In general, DL depends on data that is taken as the input, which is further processed in the layers [2]. It is a known fact that as the number of layers increases, the classification accuracy also increases. There exist the various number of learning models viz., autoencoder, deep belief network, convolutional neural network (CNN), and recurrent neural network. This section briefly outlines these models. The evolution of DL started in the late 1940s, wherein shallow neural networks (SNN) is introduced. Such networks are primitive and consist of only one hidden, input, and output layers. Figure 11.2 shows the basic SNN.

With the advancements of having K-means in the 1960s, multilayered NN and Backpropagation algorithms from the K-means era to 1970s have given wave into various researches that are being conducted to date to deep forest-based algorithms. It can be understood from Figure 11.2 that the process involved in DL is pretty simple, wherein the network is stimulated at the initial stage by input in the starting layer, which is depicted as the input variables in this case as x_1 and x_2. These are then spread to the end or output layer along with weighted connections that give out the forecast or result of restoration, which shown as $\hat{y}$. In the backward stage, connections with the weights that are defined are fine-tuned in such a way that by reducing the difference between reconstructed and original data that is provided as input.

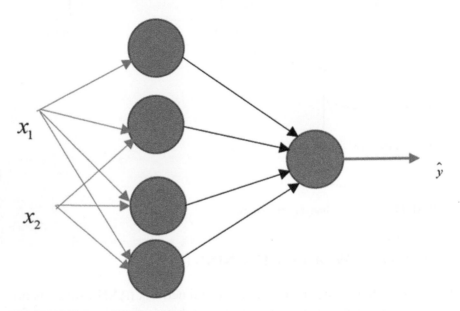

FIGURE 11.2   Shallow neural network.

The activation function forms the layers, wherein several types exist, viz., sigmoid function, ReLu, etc., [1]. A learning task can be called an optimization problem that transforms the minima of the objective function attained by proper selection of parameters. Based on the input, i.e., the types of input, the architecture and objective function needs to be selected, else will overfit the function resulting in a high error rate for the given test data that fits appropriately for the trained data.

11.3   AUTOENCODER (AE)

Autoencoder (AE) is a dimensionality reduction technique that uses compressed representation learning and minimizes the reconstruction error

[11]. Figure 11.3 shows the architecture of a basic AE. Basically, encoder, and decoder are the two components of AE. The encoder converts the input into a veiled representation, which is called a feature code. This code is formed using a mapping function. Such a mapping function is given by:

$$output = f(weight \times input + bias)$$

The output is a function of the weight multiplied by input with an additive bias. The main rule of AE is to reduce reconstruction error. AE, when combined with CNNs, uses the property of convolution to the inputs [12]. Sascha Lange and Martin Riedmiller [13] discussed the effectiveness of deep autoencoder neural networks. The main problem with AE is the zero-reconstruction error because the system learns the identity function; hence, various types of autoencoders are introduced based on the application [14]. In this context, Sufang Zhang [15] has discussed seven different types of autoencoders. They are convolutional, extreme learning machine, sparse, variational, and adversarial. All these AE's are implemented on a small-sized dataset [15]. The main advantage of AE is that new features that represent the original data can be obtained by having a simple stack-based AE. Hence, stack all the hidden layers data, resulting in the representation of the original data and then compute the reconstruction error [16, 17].

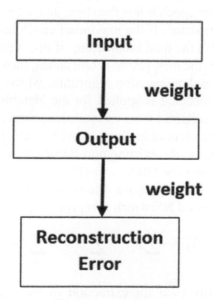

FIGURE 11.3   The architecture of basic autoencoder.

11.4   DEEP BELIEF NETWORK

Recently, deep belief networks (DBN) has gained prominence and is widely used in machine learning algorithms. DBN works on learning layer-by-layer principle. It consists of a visible layer and multiples of hidden layers. This is similar to that of the restricted Boltzmann machine (RBM). Figure 11.4 shows the basic architecture of the DBN.

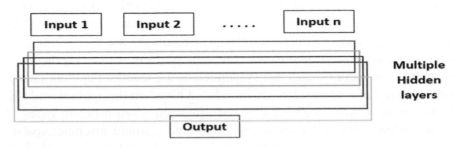

FIGURE 11.4   The architecture of deep belief network.

DBN starts with the initialization of data with maximum epoch and initial-izing the weights and biases. The next phase is to have a learning mechanism through the sigmoidal function, which is divided into positive and negative phases. If the maximum epoch is less than five, then momentum is initialized, and the weights are updated. If the maximum epoch has reached, then the momentum achieved is the final momentum. If not, a positive bias is again calculated. Further, as the momentum is calculated, then the final classifica-tion is done by the backpropagation algorithm, which itself is supervised classification. This procedure is applied for the identification of the health state. The data that is collected from various sensors over the sensor network is used, wherein the data is principally processed for effective and continuous monitoring. In contrast to the existing non-linear classification models viz., SVM, it is observed that the DBN outperformed in terms of classification rate [18]. It is known from the above discussion that DBN is built on RBM which are an energy-based NN which is represented as [22]:

$$E(v,h) = -\sum_{i \in V} a_i v_i - \sum_{j \in H} b_j h_j - \sum_{i,j} v_i h_j w_{ij}$$

These are primarily used for extraction of the dormant features that results in better classification. When combined with supervised learning,

RBM proves to have an elevated level of parallelism and rugged depiction capability [19]. This is a probabilistic generative model wherein each layer is independent of the values in other layers. The main important property of RBM is the lack of unswerving relations within the same layers but having connections between hidden layers. One can still enhance research in this area [20]. Training of DBN can be enhanced for the production of features that can be understood in a proper way, having more discernment capability by having sparsification. Using this, one can have low-level feature representation for unknowns [21]. DBN maps the given raw data into a low dimensional space by unsupervised learning. The supervised learning uses fine-tuning of the weights in the network [22]. Qili Chen et al. [23] have proposed continuous DBN (cDBN) by exploiting weak learning. cDBN consists of visible layers on the front end and output at the tail, providing labeled data that have got values transferred from hidden to visible layers, making them discrete to continuous. The two hidden layers in between have duplex connectivity. It is observed that the proposed method performs poorly when a small data set is used due to overfitting [23]. The ability of DBN is to learn in such a way that it can recover deep-categorized representation of the training data [24–26].

11.5 CONVOLUTIONAL NEURAL NETWORK (CNN)

Convolutional neural networks (CNN) has evolved its existence from the past two decades in the field of machine learning, paving its way from image processing to the latest android-based applications. The main benefit of CNN is the reduction in the number of parameters than its peers. Features that are provided as input for CNN need not be spatially dependent, which is the central assumption [27]. The main component of CNN is the convolutional layer, which learns features from the inputs. To obtain a new feature, convolving input feature maps with a kernel is performed, and results are disseminated into a non-linear function. The next layer is the pooling layer, which is having a process equivalent to that of the fuzzy filtering. This affects the second feature extraction by having dimensionality reduction with an increase in robustness. These layers generally lie between two convolutional layers. Most of the research can be concentrated in these two layers of convolutional and pooling layers, as high-level characteristics of inputs can be obtained by merely having a stack-based architecture of these two layers. The third layer is the fully connected layer (FC layer); in general, CNN has one or more FC layers with having an output layer connected to the FC layer. The classifier is programmed to be having its classification process in this layer [28]. Therefore, it can be concluded that CNN takes input, viz., images then convolve with kernels or filters. The result is used for the extraction of features [29]. Figure 11.5 shows the architecture of CNN.

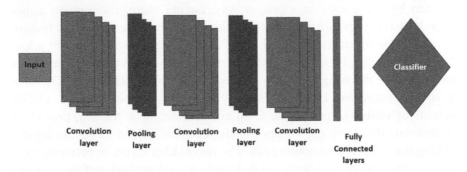

FIGURE 11.5   CNN architecture.

Training a CNN with an increase in the number of layers is a cumbersome task and requires more complex computations. This is also valid if the data sets are large in size. The last layer of CNN is a classifier. Based on the application, the classifier needs to be selected. SVM or any of the probabilistic classifiers can be used [30]. In fact, classifiers can also be considered as a fully connected dense layer. If a FC layer assists in mapping the features that are learned into a markup space, then all the other layers can be of in vain. The loss function is that it is a part of the last layers of the CNN is used for the reconstruction error prediction. This tells how well the model performs [31]. Yann [32] has presented various applications using CNN in conjunction with DL. These include starting with the representations of given image in a hierarchical way, then train the classifier based on the features that are extracted, viz., low, mid, and high-level features. DL with structure prediction, character recognition, faces detection, and estimation of the pose, etc. To improve the fastness in training or to accelerate, the rectified linear unit (ReLU) is used, and CNN updates the weighting filter using stochastic gradient descent (SGD). The performance of CNN can be improved by assigning weights for each layer, which is a heuristic process, and then follow a standard process [33]. One of the applications of the use of CNN is proposed in image classification. The network structure can be modified in such a way that it can learn and further extract the relevant features that are required, resulting in less computational complexity [34]. From the above discussion, one can conclude that CNN's are driven by the data or input, which is considered as a representative [35]. The activation functions, when are adaptive, will improve the accuracy in the prediction. This has a better application in the CT image classification [36].

11.6 RECURRENT NEURAL NETWORK (RNN)

The collection of connected nodes is called NN, wherein the nodes are nothing but neurons. The basic NN consists of the input layer, hidden layer, and output layer. Exploiting the drawback of NN, which predicts the output of fixed size based on the input, which is also of fixed size, recurrent neural network (RNN) is proposed as a solution. RNN works in a raster scan way. The functionality of RNN depends on the present and past inputs [37]. Figure 11.6 shows the architecture of RNN.

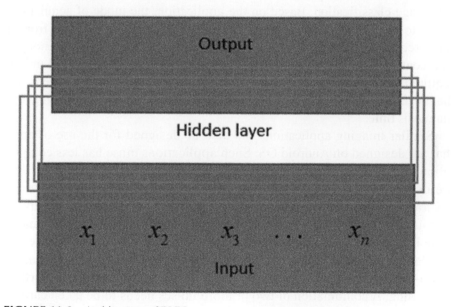

FIGURE 11.6   Architecture of RNN.

RNN has extended its applications to 5G communications. Turbo codes are used for error correction wherein RNN architecture is used in comparison to the existing conventional turbo coders [38]. One more application of RNN with DL is in speech processing. Automatic speech recognition is the main task of any speech processing system with ML. RNN is trained to handle this problem [39]. The final layers output is calculated as the average of the output of the two parallel recurrent layers, which is used for image classification. Such a technique is observed to have better accuracy with real-time data [40–42].

11.7   APPLICATIONS OF DEEP LEARNING (DL)

Health monitoring is a prime challenge in the latest scenario that is prevalent across the world. Diagnosis, along with prognosis, is the prima facie that is keeping ahead in the health diagnosis, which is the prime requirement of any individual. Health diagnosis is made from rapid test kits to long end testing for instant identification to perfect detection. At this juncture, some ailments require imaging for proper diagnosis. Examples include the detection of diabetic retinopathy (DR), cancer identification, fracture detection, CT image classification. Based on the application, the mode of acquisition needs to be decided. DR requires fundus images that are acquired from a fundoscopy, whereas for fracture detection, only x-ray images are enough. In recent days, a hard copy of the x-ray images is not provided until unless required. The acquired images are sent directly to the doctor, where the diagnosis is carried out in the system itself, reducing the cost of printing and diagnosis time.

Similar imaging applications can also be designed for the use of apps that are designed on Android OS. Such applications muse has less computational complexity when compared to the heavyweight applications that are deployed on a stand-alone system with other types of OS. There are several steps from acquisition to diagnosis. This section clearly discusses various applications pertaining to health. Image enhancement is normally done by the operator, who enhances the image that is acquired. However, standardization needs to be done before storing it in the DB. This is called a pre-processing stage, where standardization is completed for further processing. A combination of two or more noises needs to be taken care of while acquisition. Such noise may reduce the quality of the image. Normal models cannot reduce the noise and restore the image. Hence, a novel technique is proposed using which better noise reduction is achieved in contrast to the existing techniques [43].

11.7.1   DETECTION OF DIABETIC RETINOPATHY (DR)

Diabetes, a disorder caused due to the malfunction of the pancreas. Such a disorder causes long-term ailments to the humans, viz., retinopathy, which is termed as DR. Diabetes, in turn, affects blood vessels throughout the body, eyes, kidneys, and joints. An estimate of the records reveals that people suffering from DR may rise to approx. of 191 million in the coming decade, which is an estimated value based on the currently available statistics. The result of having such an ailment leads to a tremendous effect on the vision, creating a disturbance in the vision capability and often leading to complete blindness. If treated in time, loss of vision can be reduced to a certain extent [44]. Hence, there is a need of the hour for early detection of DR, which has long-term effects. At this juncture, it is also appropriate to grade DR, hence, deriving the stage at which the ailment is. Patients may not have symptoms in the initial stages of DR; while in later stages will lead to blood leakage in the eye. The signs of DR are exudates and micro-aneurysms, wherein the next stage is the hemorrhage appearance [45, 46]. These are called a pathological sign which is visible through medical imaging [47]. Figure 11.7 shows a normal eye and DR affected eye. Exudates are of two types, soft, and hard. The former and the latter can be viewed as a bright yellow lesion. The severity of the disease can be assessed by the spatial distribution of microaneurysms, exudates, and hemorrhages in relation to fovea [48]. It is observed that some of the persons suffering from this ailment reported having floating dark spots as well as a distorted image. Such a distorted vision is due to the leakage of the blood from damaged vessels of the eye [49]. Hence, it can be concluded that early detection may help patients suffering from such disease to get better treatment. The automatic identification of the lesions and fundus image abnormalities will help ophthalmologists for efficient prognosis [50]. The categorization of DR can be performed based on the spatial distribution of the components of the image and is classified as mild, moderate, severe, and advanced [51, 52]. In order to detect the stage of the DR and for prognosis, features need to be extracted from the eye, i.e., retinal image. The features that are extracted from the retinal images must pave the way for better diagnosis. Hence, better feature extraction techniques need to be considered. The next clearly outline the feature extraction process.

11.7.2   FEATURE EXTRACTION AND CLASSIFICATION

Feature extraction is a crucial step in image classification. If the features that are representing an image are perfectly chosen, then the classification process will be easy. Hence, the features that are selected should be selected appropriately. Features can be of two types, basically. They are low-level and high-level features. The former poses a problem to the classifier as all the features will be of the same domain or will be in the same. Hence the classification process becomes cumbersome as the DB size increases. Hence, features will be transformed into a higher plane, or high-level features will be extracted. Color, texture, and some of the statistical features that are extracted from similar images will be close to each other. Hence, the classification process will result in wrong class detection, or a wrong class will be identified.

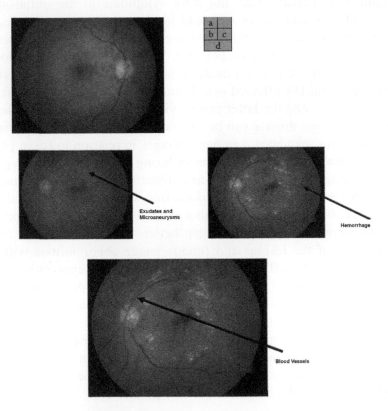

FIGURE 11.7   Normal and DR affected eye. (a) Normal Fundus image. (b) Exudates and microaneurysms. (c) Hemorrhage in eye. (d) Blood vessels in the eye.

High-level features are obtained by projecting low-level features using PCA, ICA, or using kernels. A kernel is a function that projects the given data into an n-dimensional subspace making non-linear classification into a linear one, thus reducing the computational complexity [57]. Features like entropy, standard deviation, mean, correlation, homogeneity, Euler number, and contrast are computed using a discrete cosine transform (DCT). These features are further classified using multi-layer perceptron neural network (MLPNN) and achieved an accuracy of 98% within a limited dataset [44]. The texture is another low-level feature that is used predominantly in image classification. The texture is nothing but a repeated pattern that is typically extracted from the images. If Figure 11.7 is clearly observed, the eye image consists of a textured pattern. Hence, local ternary pattern (LTP) and local energy-based shape histogram (LESH) [45] are extracted from the retinal image. SVM is used as a classifier and achieved an accuracy of 90% by the combined feature of LTP and LESH. CNN, along with the monocular and binocular model, has achieved an accuracy of 94% and 95.1%, respectively [46]. A review of various types of features that are extracted from the retinal images is provided, and most of the features presented are low-level features [47]. As the images that are acquired through fundoscopy are of color in nature, color models-based classification can be performed. However, the segmentation process must be accurate so that proper color analysis can be done. Segmentation is performed using K-means clustering, and different parts of the image with DR are identified viz., hemorrhages, hard, and soft exudates, etc., [48]. Morphology [51] is the study of structure. Such a study, when applied to retinal images, observed to provide better results. A hole entropy enabled decision tree classifier along with hybrid features [50] is proposed for the extraction of exudates from the retinal images. All these techniques try to provide better analysis and representation of the ailment in terms of exudates, hemorrhages.

The next step is a classification, which is crucial in the identification of retinopathy. Figure 11.10 shows the classification accuracy achieved by various techniques. It can be observed that NN based techniques have achieved better results. These NN-based techniques, when combined with DL, achieved better accuracy in classification. Researchers can concentrate on increasing the number of layers that provide better accuracy at the cost of an increase in computational complexity. Hence, there is a need of the hour to optimize the features and layers. However, in such cases, the accuracy decreases. Therefore, a balance mechanism needs to be experimented for further analysis to improve the performance of the system. Computed tomography (CT) is a technique that combines x-ray images acquired from different angles and poses. Thus, the formed cross-sectional slices of the bones, blood vessels, soft tissues are being processed in the computer for further analysis. Such an analysis provides a better prognosis than x-ray imaging. Hence, CT is more preferred when the ailment is not possible for identification via x-ray imaging. In this juncture, lung CT imaging clas-sification has been discussed in this section. Interstitial lung disease (ILD) consists of more than 150 diseases that affect the interstitium, wherein the ability to breathe gets impaired. The classification rate achieved is 95% when CNN is applied [35]. Another application of CT imaging is in brain image classification. Figure 11.8 shows a CT image of the brain [56]. The classification of any tumor present in the brain as benign or malignant is based on features extracted. DL has made the task easy, hence, making automatic or automated classification when CT images are given as input (Figures 11.8–11.10).

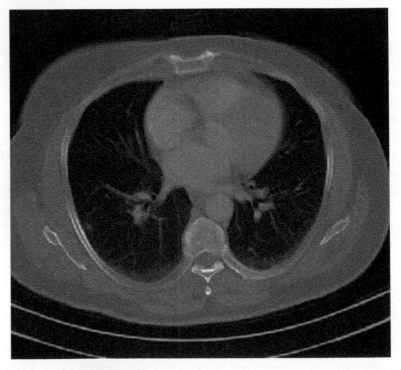

FIGURE 11.8 Brain image slice.
Source: Courtesy: Kaggle Database.

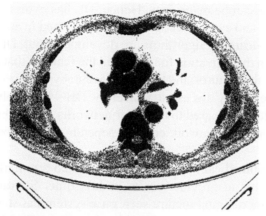

FIGURE 11.9   2D image view of brain CT image.
Source: Courtesy: Kaggle Database.

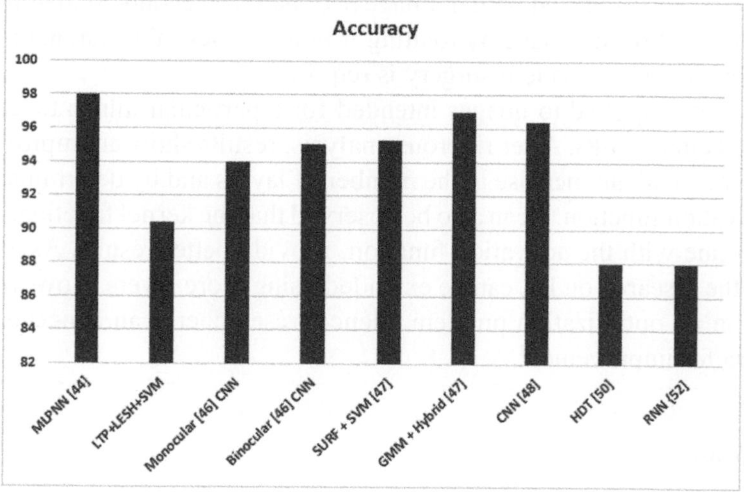

FIGURE 11.10   Classification accuracy of different techniques.

11.8 CONCLUSION

The ailments that are progressing with unseen organisms are increasing day by day. Prognosis of the ailments needs to be developed in an exponential manner, thus making early detection available. Features from imaging or symptoms of the ailment need to be tabulated for further analysis. As more number of features that can predict the stage of the ailment is available, early detection process can be made easy. Hence, DL has evolved that can make this process easy, and early detection can be performed in an automated way. Various machine-learning algorithms are developed using DL. The accuracy of DL depends on the activation function and the features that are provided as input. The more the features, the better the accuracy. In this chapter, various blocks of DL are clearly discussed. Currently, DBN has evolved and proved to be the best when compared to the other networks. It may be noted that the accuracy of a particular network entirely depends on the type of application and the features.

Imaging in biomedical applications has marked a remarkable response, as it made systems to provide results in less time. Starting from analog imaging to having a digital picture seen on a system has made us view the images of the human body on screen with a better visualization. This helped doctors to have a better view so that they can enhance or move to the location as required. The advantage of having such visualization can be seen in two-fold. One in reduced time in image acquisition. Second, in identifying a particular part of the image by rotating it in a 3D view, thus having a better mark for a cut to be made if surgery is required.

DL, when applied to images intended for a particular ailment, seems to provide better results. After rigorous analysis, results show an improvement in accuracy with an increase in the number of layers and by determination of the activation function. It can also be observed that the kernel functions, when used in line with the activation function, provide better results. As a future scope, the research on DL can be extended using more layers. However, this results in an optimization problem. Hence, researchers can concentrate on this area for improvement.

KEYWORDS

- **autoencoder**
- **computed tomography**
- **convolutional neural networks**
- **deep belief networks**
- **diabetic retinopathy**
- **discrete cosine transform**
- **interstitial lung disease**
- **local energy-based shape histogram**

REFERENCES

1. Chensi, C., Feng, L., Hai, T., Deshou, S., Wenjie, S., Weizhong, L., Yiming, Z., et al., (2018). Deep learning and its applications in biomedicine. *Genomics Proteomics Bioinformatics, 16,* 17–32.
2. Xuedan, D., Yinghao, C., Shuo, W., & Leijie, Z., (2016). Overview of deep learning. In: *31st Youth Academic Annual Conference of Chinese Association of Automation.* Wuhan, China.
3. Xinyi, Z., Wei, G., Wen, L. F., & Fengtong, D., (2017). application of deep learning in object detection. In: *2017 International Conference on Computer and Information Science.* Wuhan, China.
4. Tamer, K., Uz. Selim, S., Op. Dr. Gökhan, Ç., & Ali, O., (2019). Interpretable machine learning in healthcare through generalized additive model with pairwise interactions (GA2M): Predicting severe retinopathy of prematurity. In: *2019 International Conference on Deep Learning and Machine Learning in Emerging Applications (Deep-ML)* (pp. 61–66).
5. Lee, K. B., & Shin, H. S., (2019). An application of a deep learning algorithm for automatic detection of unexpected accidents under bad CCTV monitoring conditions in tunnels. In: *2019 International Conference on Deep Learning and Machine Learning in Emerging Applications (Deep-ML)* (pp. 7–11) Istanbul, Turkey.
6. Halil, C. K., Sezer, G., (2019). A deep learning based distributed smart surveillance architecture using edge and cloud computing. In: *2019 International Conference on Deep Learning and Machine Learning in Emerging Applications (Deep-ML)* (pp. 1–6). IEEE.
7. Mohamed, A. A., (2018). *Improving Deep Learning Performance Using Random Forest HTM Cortical Learning Algorithm* (pp. 13–18). IEEE.
8. Ochin, S., (2019). Deep challenges associated with deep learning. In: *2019 International Conference on Machine Learning, Big Data, Cloud and Parallel Computing (Com-IT-Con)* (pp. 72–75). India.
9. Arshiya, B., Farheen, F., & Asfia, S., (2019). Implementation of deep learning algorithm with perceptron using tensor flow library. *International Conference on Communication and Signal Processing,* 172–175. India.
10. Alexander, S. L., & Arvid, L., (2019). An overview of deep learning in medical imaging focusing on MRI. *Zeitschrift für Medizinische Physik, 29*(2), 102–127.
11. Wang, W., Huang, Y., Wang, Y., & Wang, L., (2014). Generalized autoencoder: A neural network framework for dimensionality reduction. In: *2014 IEEE Conference on Computer Vision and Pattern Recognition Workshops* (pp. 496–503). Columbus, OH.
12. Hou, B., & Yan, R., (2018). Convolutional auto-encoder based deep feature learning for finger-vein verification. In: *2018 IEEE International Symposium on Medical Measurements and Applications (MeMeA)* (pp. 1–5). Rome.
13. Lange, S., & Riedmiller, M., (2010). Deep auto-encoder neural networks in reinforcement learning. In: *2010 International Joint Conference on Neural Networks (IJCNN)* (pp. 1–8). Barcelona.
14. Jiang, X., Zhang, Y., Zhang, W., & Xiao, X., (2013). A novel sparse auto-encoder for deep unsupervised learning. In: *2013 Sixth International Conference on Advanced Computational Intelligence (ICACI)* (pp. 256–261). Hangzhou.

15. Zhai, J., Zhang, S., Chen, J., & He, Q., (2018). Autoencoder and its various variants. In: *2018 IEEE International Conference on Systems, Man, and Cybernetics (SMC)* (pp. 415–419). Miyazaki, Japan.

16. Meng, Q., Catchpoole, D., Skillicom, D., & Kennedy, P. J., (2017). Relational autoencoder for feature extraction. In: *2017 International Joint Conference on Neural Networks (IJCNN)* (pp. 364–371). Anchorage, AK.

17. Chu, H., Xing, X., Meng, Z., & Jia, Z., (2019). Towards a deep learning autoencoder algorithm for collaborative filtering recommendation. In: *2019 34ᵗʰ Youth Academic Annual Conference of Chinese Association of Automation (YAC)* (pp. 239–243). Jinzhou, China.

18. Tamilselvan, P., Wang, Y., & Wang, P., (2012). Deep belief network-based state classification for structural health diagnosis. In: *2012 IEEE Aerospace Conference* (pp. 1–11). Big Sky, MT.

19. Ye, Z., et al., (2019). Learning parameters in deep belief networks through ant lion optimization algorithm. In: *2019 10ᵗʰ IEEE International Conference on Intelligent Data Acquisition and Advanced Computing Systems: Technology and Applications (IDAACS)* (pp. 548–551). Metz, France.

20. Gao, N., Gao, L., Gao, Q., & Wang, H., (2014). An Intrusion detection model based on deep belief networks. In: *2014 Second International Conference on Advanced Cloud and Big Data* (pp. 247–252). Huangshan.

21. Keyvanrad, M. A., & Homayounpour, M. M., (2015). Normal sparse deep belief network. In: *2015 International Joint Conference on Neural Networks (IJCNN)* (pp. 1–7). Killarney.

22. Zhao, G., Zhang, C., & Zheng, L., (2017). Intrusion detection using deep belief network and probabilistic neural network. In: *2017 IEEE International Conference on Computational Science and Engineering (CSE) and IEEE International Conference on Embedded and Ubiquitous Computing (EUC)* (pp. 639–642). Guangzhou.

23. Chen, Q., Pan, G., Qiao, J., & Yu, M., (2019). Research on a continuous deep belief network for feature learning of time series prediction. In: *2019 Chinese Control and Decision Conference (CCDC)* (pp. 5977–5983). Nanchang, China.

24. Skaria, S., Mathew, T., & Anjali, C., (2017). An efficient image categorization approach using deep belief network. In: *2017 International Conference on Networks and Advances in Computational Technologies (NetACT)* (pp. 9–14). Thiruvanthapuram.

25. Rani, R. D. K. G., & Mahendra, C. R., (2019). Eye disease classification based on deep belief networks. *International Journal of Recent Technology and Engineering (IJRTE)*, *8*(2S11), 3273–3278.

26. Yuming, H., Junhai, G., & Hua, Z., (2015). Deep belief networks and deep learning. *Proceedings of 2015 International Conference on Intelligent Computing and Internet of Things*, 1–4. Harbin.

27. Albawi, S., Mohammed, T. A., & Al-Zawi, S., (2017). Understanding of a convolutional neural network. In: *2017 International Conference on Engineering and Technology (ICET)* (pp. 1–6). Antalya.

28. Guo, T., Dong, J., Li, H., & Gao, Y., (2017). Simple convolutional neural network on image classification. In: *2017 IEEE 2ⁿᵈ International Conference on Big Data Analysis (ICBDA)* (pp. 721–724). Beijing.

29. Chauhan, R., Ghanshala, K. K., & Joshi, R. C., (2018). Convolutional neural network (CNN) for image detection and recognition. In: *2018 First International Conference*

on Secure Cyber Computing and Communication (ICSCCC) (pp. 278–282). Jalandhar, India.

30. Aloysius, N., & Geetha, M., (2017). A review on deep convolutional neural networks. In: *2017 International Conference on Communication and Signal Processing (ICCSP)* (pp. 0588–0592). Chennai.

31. Yang, J., & Li, J., (2017). Application of deep convolution neural network. In: *2017 14th International Computer Conference on Wavelet Active Media Technology and Information Processing (ICCWAMTIP)* (pp. 229–232). Chengdu.

32. LeCun, Y., (2015). Deep learning and convolutional networks. In: *2015 IEEE Hot Chips 27 Symposium (HCS)* (pp. 1–95). Cupertino, CA.

33. Hayat, S., Kun, S., Tengtao, Z., Yu, Y., Tu, T., & Du, Y., (2018). A deep learning framework using convolutional neural network for multi-class object recognition. In: *2018 IEEE 3rd International Conference on Image, Vision and Computing (ICIVC)* (pp. 194–198). Chongqing.

34. Al-Saffar, A., A. M., Tao, H., & Talab, M. A., (2017). Review of deep convolution neural network in image classification. In: *2017 International Conference on Radar, Antenna, Microwave, Electronics, and Telecommunications (ICRAMET)* (pp. 26–31). Jakarta.

35. Shin, H., et al., (2016). Deep convolutional neural networks for computer-aided detection: CNN architectures, dataset characteristics and transfer learning. In: *IEEE Transactions on Medical Imaging* (Vol. 35, No. 5, pp. 1285–1298).

36. Zahedinasab, R., & Mohseni, H., (2018). Using deep convolutional neural networks with adaptive activation functions for medical CT brain image classification. In: *2018 25th National and 3rd International Iranian Conference on Biomedical Engineering (ICBME)* (pp. 1–6). Qom, Iran.

37. Kaur, M., & Mohta, A., (2019). A review of deep learning with recurrent neural network. In: *2019 International Conference on Smart Systems and Inventive Technology (ICSSIT)* (pp. 460–465). Tirunelveli, India.

38. Sattiraju, R., Weinand, A., & Schotten, H. D., (2018). Performance analysis of deep learning based on recurrent neural networks for channel coding. In: *2018 IEEE International Conference on Advanced Networks and Telecommunications Systems (ANTS)* (pp. 1–6). Indore, India.

39. Tang, Z., Wang, D., & Zhang, Z., (2016). Recurrent neural network training with dark knowledge transfer. In: *2016 IEEE International Conference on Acoustics, Speech and Signal Processing (ICASSP)* (pp. 5900–5904). Shanghai.

40. Chandra, B., & Sharma, R. K., (2017). On improving recurrent neural network for image classification. In: *2017 International Joint Conference on Neural Networks (IJCNN)* (pp. 1904–1907). Anchorage, AK.

41. Baktha, K., & Tripathy, B. K., (2017). Investigation of recurrent neural networks in the field of sentiment analysis. In: *2017 International Conference on Communication and Signal Processing (ICCSP)* (pp. 2047–2050). Chennai.

42. Abroyan, N., (2017). Convolutional and recurrent neural networks for real-time data classification. In: *2017 Seventh International Conference on Innovative Computing Technology (INTECH)* (pp. 42–45). Luton.

43. Satyanarayana, T. V. V., Alivelu, M. N., Pradeep, K. G. V., & Venkata, N. M., (2020). Mixed image denoising using weighted coding and non-local similarity. *SN Appl. Sci.*

44. Bhatkar, A. P., & Kharat, G. U., (2015). Detection of diabetic retinopathy in retinal images using MLP classifier. In: *2015 IEEE International Symposium on Nanoelectronic and Information Systems* (pp. 331–335). Indore.

45. Chetoui, M., Akhloufi, M. A., & Kardouchi, M., (2018). Diabetic retinopathy detection using machine learning and texture features. In: *2018 IEEE Canadian Conference on Electrical and Computer Engineering (CCECE)* (pp. 1–4). Quebec City, QC.

46. Zeng, X., Chen, H., Luo, Y., & Ye, W., (2019). Automated diabetic retinopathy detection based on binocular Siamese-like convolutional neural network. In: *IEEE Access* (Vol. 7, pp. 30744–30753).

47. Ahmad, A., Mansoor, A. B., Mumtaz, R., Khan, M., & Mirza, S. H., (2014). Image processing and classification in diabetic retinopathy: A review. In: *2014 5th European Workshop on Visual Information Processing (EUVIP)* (pp. 1–6). Paris.

48. ManojKumar, S. B., & Sheshadri, H. S., (2016). Classification and detection of diabetic retinopathy using K-means algorithm. In: *2016 International Conference on Electrical, Electronics, and Optimization Techniques (ICEEOT)* (pp. 326–331). Chennai.

49. Kajan, S., Goga, J., Lacko, K., & Pavlovičová, J., (2020). Detection of diabetic retinopathy using pretrained deep neural networks. In: *2020 Cybernetics and Informatics (K&I)* (pp. 1–5). Velke Karlovice, Czech Republic.

50. Shirbahadurkar, S. D., Mane, V. M., & Jadhav, D. V., (2017). A modern screening approach for detection of diabetic retinopathy. In: *2017 2nd International Conference on Man and Machine Interfacing (MAMI)* (pp. 1–6). Bhubaneswar.

51. Omar, Z. A., Hanafi, M., Mashohor, S., Mahfudz, N. F. M., & Muna'im, M., (2017). Automatic diabetic retinopathy detection and classification system. In: *2017 7th IEEE International Conference on System Engineering and Technology (ICSET)* (pp. 162–166). Shah Alam.

52. Kanungo, Y. S., Srinivasan, B., & Choudhary, S., (2017). Detecting diabetic retinopathy using deep learning. In: *2017 2nd IEEE International Conference on Recent Trends in Electronics, Information and Communication Technology (RTEICT)* (pp. 801–804). Bangalore.

53. Ankita, G., & Rita, C., (2018). Diabetic retinopathy: Present and past. *Procedia Computer Science, 132,* 1432–1440.

54. Kauppi, T., Kalesnykiene, V., Kamarainen, J. K., Lensu, L., Sorri, I., Raninen, A., Voutilainen, R., et al. (2007). *DIARETDB1 Diabetic Retinopathy Database and Evaluation Protocol.* Technical report (PDF).

55. Kauppi, T., Kalesnykiene, V., Kamarainen, J. K., Lensu, L., Sorri, I., Raninen, A., Voutilainen, R., et al., (2017). DiaRetDB1 diabetic retinopathy database and evaluation protocol. In: *Proc. of the 11th Conf. on Medical Image Understanding and Analysis.* (Aberystwyth, Wales). Accepted for publication.

56. Wang, X., Peng, Y., Lu, L., Lu, Z., Bagheri, M., & Summers, R. M., (2017). *ChestX-ray8: Hospital-Scale Chest X-Ray Database and Benchmarks on Weakly-Supervised Classification and Localization of Common Thorax Diseases.* IEEE CVPR.

57. Tallapragada, V. V. S., & Rajan, E. G., (2012). Improved kernel-based IRIS recognition system in the framework of support vector machine and hidden Markov model. *IET Image Processing, 6*(6), 661–667.

INDEX

Printed and bound by CPI Group (UK) Ltd, Croydon, CR0 4YY

23/10/2024

01777702-0013